I0045148

A. Wellington Adams

Electricity

Its Application in Medicine and Surgery. Vol. 1

A. Wellington Adams

Electricity
Its Application in Medicine and Surgery. Vol. 1

ISBN/EAN: 9783337779108

Printed in Europe, USA, Canada, Australia, Japan

Cover: Foto ©berggeist007 / pixelio.de

More available books at **www.hansebooks.com**

A Modern "Night-Call."

ELECTRICITY

Its Application in Medicine and Surgery.

A Brief and Practical Exposition of Modern Scientific Electro-Therapeutics.

— BY —

WELLINGTON ADAMS, M. D.,

*Author of "Art of Telephony—By Whom Discovered;" "Evolution of the
Electric Railway;" "Design and Construction of Dynamo-Electric
and Electro-Dynamic Machinery;" Lecturer on Electro-
Therapeutics, University Medical College, Kansas
City; Formerly Professor of Diseases of the
Ear, Nose, and Throat, Medical De-
partment, University of Den-
ver, and Editor "Rocky
Mountain Medical
Review."*

VOLUME I.

1891.
GEORGE S. DAVIS,
DETROIT, MICH.

This little book is, by permission, respectfully dedicated to

FRANCIS E. NIPHER,

*Professor of Physics in Washington University, whose unsel-
fish devotion to the cause of science has commanded
the author's admiration and respect.*

CHAPTER I.

INTRODUCTION.

The writer, having, without doubt, devoted an unusual amount of time and study, not only to electro-therapeutics, but also to original experimental research in general electrical science, rendering himself thoroughly conversant with electrical engineering in all its branches, even to such an extent as to have introduced the electric railway and the electrical transmission of power into this country, as well as assisted in familiarizing electrical engineers with the principles involved in the scientific design and construction of commercial electro-dynamic and dynamo-electric machinery, will perhaps be pardoned for assuming what to many will appear a somewhat dogmatic attitude in this brief exposition of the science of electrotherapeutics. In dealing with a subject covering so much ground, with the brevity that the confines of the present work demands, one must, however, speak to the point and authoritatively, avoiding all lengthy discussions and descriptions of the highways and byways leading to the conclusions enunciated. Should the statements of the writer be questioned, recognized authorities must be consulted.

The "medical electrician," and the average "electro-therapeutic text-book," have long since become the butt of amusement and laughing-stock of the physicist and electrical engineer. This condition of affairs is fully merited, and results from the unscientific methods of discussion and investigation that have thus far characterized the crude development of this most important, but sadly neglected branch of legitimate medicine. The most fallacious ideas have prevailed concerning the nature of electricity and the laws governing its manifestations. The failure to use proper instruments of precision has given rise to the most contradictory reports from the same and different operators as to the merits of electricity in the various pathological conditions which come under the observation of the physician. The rank and file of the profession who have undertaken to use it as a remedial agent have paid no attention whatever to the *technique* of the science, and the same may even be said of many of those savants who have made some considerable study of electro-therapeutics. To undertake the practice of electro-therapeutics without a thorough knowledge of the fundamental principles and laws of the science of electricity and magnetism, is as ridiculous and impracticable as would be an effort to carry on chemical analyses without having first become conversant with the principles of chemistry. In no science is a knowledge of general principles and natural laws so essential at every step as in the case of electrical science. Its laws are multiple, varied and complex, and a thorough understanding of them is necessary for a correct interpretation of the numerous phenomena which are ever and anon manifesting themselves. It is no uncommon thing to see a physician dodging around with a stereotyped form of "medical battery" in his hand, proposing to destroy hair follicles by electrolytic action through the agency of a Faradic current; neither is it an unusual occurrence to hear the "medical electrician" refer to the negative and positive pole of a Faradic battery, and to speak of Static, Faradic and Galvanic electricity, as though they were all separate and distinct entities. In their conception of the various properties or qualities of an electric current, such as intensity, tension, and quantity, we find them equally at sea, and therefore constantly misapplying these terms. Many of these criticisms may be applied to some of the more modern writers upon the subject who speak very glibly about *ohms*, *volts* and *milli-amperes*. The methods of the past have been purely empirical, and therefore uncertain and varied results have been chronicled, until it is a wonder the science has any standing in legitimate medicine. The high esteem in which it is to-day held by the leading members or the medical profession, results from a forced recognition of the numerous accidental but brilliant results which have occasionally been stumbled upon, and not from any published uniformity of action. Electricity, however, is a definite quantity capable of producing with uniformity, certain physical, chemical and nutrient changes, whenever we come to understand the laws governing its action, and learn to apply it with methods of precision.

A few writers of very recent date, more particularly in the field of gynæcological electro-therapeutics, have approached nearer to a scientific exposition of the subject. I refer especially to the writings of Apostoli, Tripier, Engelmann, and Massey. Much of the existing empiricism has resulted from the absence of, and failure to use, suitable and reliable measuring instruments. Progress in every department of science is largely dependent on exact measurements, since it is only by this means that we obtain an accurate knowledge of relative values. The thermometer enables us to investigate the the laws of heat; the barometer furnishes a knowledge of atmospheric pressure and the various matters relating to it. In chemistry and astronomy, almost every step depends upon such measurement. Even our ordinary business transactions, and the value of our currency, are regulated by the common scales by which we measure the force of gravity. Electric science is no exception to this rule. We require to know, accurately, relative differences of potential and current strength; the conductivity and resistance of various substances and tissues; the force of electric attraction and repulsion; the comparative energy of the various instruments used for generating and accumulating electricity; and other matters of equal importance.

Electric measurement, however, presents peculiar difficulties not met with in the measurement of other forms of energy. In the measurement of gravity, we deal with a force easily controlled, the direction of whose movement is always known, and which, on the various points of the earth's surface, is subject to but slight variation; in heat we have an energy suscepti ble of easy control, its movement is slow, and its direction easily ascertained. Electricity, on the other hand, moves with greater rapidity than thought; its direction is difficult to ascertain, and it is not readily subservient to absolute control; so that the results of measurement, by even some of our best constructed instruments, fall short of what we would desire. Any very great inaccuracy of measurement, however, is worse than no attempt at measurement The physician who imagines that he can deal with electricity as a remedial agent and secure anything approximating uniform results, without the use of reliable measuring instruments, will find himself as much at sea as the mariner without his compass, the carpenter without his rule, or the pharmacist without his scales.

The ignorance which prevails concerning the physics of the subject can be illustrated in no better way than by citing the fact that the writer has frequently been applied to by brother practitioners who wished to know what form or kind of battery would be best for all purposes, since their means were limited and they could afford to purchase but one. Let us imagine a would-be pharmacist applying to us to know what would be the best single drug to place on sale in a prospective drug store, which would be likely to meet all cases and emergencies, in lieu of his being obliged to assume the expense of a stock comprising a varied assortment of drugs. What would be our answer to such an applicant? Must we not condole his lack of means, and advise him to enter upon the pursuit of some business more commensurate with his capital? The one proposition is no more absurd than the other.

CHAPTER II.

ELECTRICITY AND MAGNETISM.

THE CONDITION OF ELECTRIFICATION is a peculiar and unusual property which most forms of matter may assume under certain specific conditions, rendering the matter so electrified susceptible of manifesting peculiar and unusual phenomena. For instance, if two sheets or pieces of dissimilar metal, such, for example, as copper and zinc, be immersed in an acidulated solution and connected together metallically, the metallic connection will assume a property which it did not before possess, the condition it will assume being that of " electrification." In other words, what is called an " electric current " will flow through the connecting metal or through the wire, in case the connecting metal assumes that form. Under this condition, the wire is capable of manifesting many strange phenomena. For instance, it will possess the power, which it did not have before, of attracting iron filings, and, if it be placed over and parallel with a delicately suspended light magnet, such as the needle of a compass, it will cause the needle to turn on its axis and place itself at right angles to the wire. If this same wire be coiled into a spool, or " helix " as it is technically termed, such spool, when suspended freely, will assume a directive force, such as will place it in the line of the magnetic meridian of the earth, which corresponds very nearly with the geographical axis, and that end which points to the North we call the north pole of such helix, although in reality it is the south pole, because, unlike magnetic poles attract each other, while like poles repel each other. If this same spool be thrust into iron filings, a great number of such filings will collect around each end of the spool, and this marks the position of what is termed the "poles" of the helix. Such a spool will also have the power to attract to itself small pieces of iron and steel, when brought sufficiently near such substances. Again, if we insert into the middle of this spool, a piece of *soft* iron, the latter takes upon itself that property which is termed "magnetism," so long as the so-called "current of electricity" continues to flow, or, in more exact words, so long as the wire which goes to make up the spool remains in the condition of electrification; the moment the current ceases to flow, or the wire loses this peculiar property which we call electrification, the magnetism of the iron also ceases, or, as it is technically termed, the iron becomes "demagnetized." If, in place of a piece of soft iron, we were to insert into the centre of this helix, a piece of *hard* iron or steel, the latter would become permanently magnetized; that is to say, the steel will not lose its magnetism so soon as the current of electricity ceases to flow through the wire, but will become a "*permanent*" magnet. A spool of wire such as we have described with a piece of soft iron in its centre, is technically known as an "electro-magnet." Electro-magnets are to be found in all telegraph sounders, telephone receivers, electric bells, electric light generators, and, in fact, in nearly every piece of electro-mechanism.

WHAT IS ELECTRICITY; or by virtue of what peculiar state of affairs does this condition of electrification exist? This is a question which has for ages in the past challenged an answer from those most learned in the science of natural philosophy. Various theories have waxed and waned. The single- and the double-fluid hypotheses, with others, have, each in their turn had their day and coterie of adherents, only to be finally consigned to a place amongst the myths of the past. Now, however, what we call electricity has come to be looked upon by nearly all the leading scientists of the present age, as having no existence *per se*, apart and distinct from the matter which it affects; in other words, as being in no sense a distinct entity like a gas or liquid, but rather as being one of the many forms of energy, a simple mode of motion or molecular vibration, similar to light, heat, and sound, into all of which it is convertible, subject to the laws governing the correlation and conservation of physical forces. All of the various forms of energy—heat, light, sound, chemical action, and electricity—are interconvertible one with the other. Unlike sound, heat, and light, however, the molecular vibration which accompanies or gives rise to the manifestation of electric energy, assumes a certain directive force, which it also has the power of impressing upon surrounding matter. Under this conception of the nature of electricity, and other theories based upon it, have been developed the various industrial applications of electricity to their present marvelous state of perfection. In the case of electric generators and electric-light machines, for instance, we are now no longer dependent upon empirical methods, upon rules of thumb, and upon traditional practice, as in the past. It is not even sufficient now to be able to make one particular kind or type of such machine so that it will simply work, but we must also know how to construct all possible varieties with the highest degree of perfection. For a long time our text-books and current electrical literature were filled with intricate formulæ, involving integral calculus and other difficult branches of higher mathematics, all of which involved the use of certain constants which had to be predetermined for each particular machine and for each particular type of machine. There were no generalizations, no universally applicable formulæ, no general constants, so that these formulæ were practically useless, and the electrical engineer was compelled to resort to rules of thumb and traditional practices — much of the same state of affairs existing as we now find in electro-therapeutics. To-day, however, it is quite different. We can now with mathematical precision calculate the best size, shape, and exciting power of a pair of " field magnets " for such machines, and the exact speed at which a given machine must be run to produce a certain electric output. To do this, however, we are obliged to resort to other laws and to other theories than those enunciated by Lenz, Jacobi, Dub, Muller, and others, in the past, concerning magnetism.

For those formulæ and theories upon which we rely to-day, we are indebted to Frohlich, Deprez, Thomson, Hopkinson and Kapp, who, by their conceptions and theories, have so unravelled the mysteries and fundamental principles of dynamic electricity, as to render it possible at this day to construct an almost theoretically perfect piece of electric apparatus. The present state of this art is due almost entirely to that conception which looks upon both an electric current and a magnet as emitting what are called " *lines of force*," which, in their passage through different media, meet with a " *resistance* " proportionate to, and dependent upon, the character of the matter they traverse. Whether such lines do really exist or not, is of but little moment for our purpose, since it suffices to know that by means of iron filings we can make such lines visible and study their properties. According to this conception, every wire carrying a current of electricity is surrounded by a multitude of circular lines arranged like concentric rings around the wire, thus (see Fig. 1), which lines gradually diminish in number as the distance from the wire increases. If we could have a *single* magnetic pole, that is, a piece of steel having only one

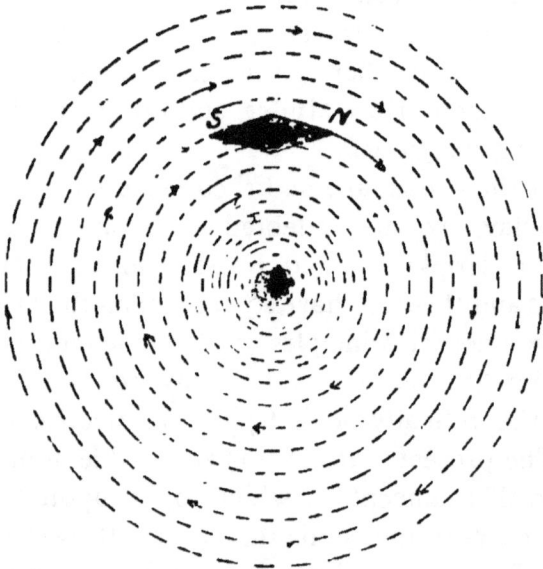

FIG. 1.

pole, and that a north pole, for instance, this pole would, if brought near a wire carying a current of electricity, revolve around the wire; but since a magnet with only one pole is an impossibility, and since the same force which urges the north pole in one direction, would urge the south pole in the opposite direction, the magnet as a whole cannot revolve

around a straight wire, but will set itself, as already
stated, at right angles to the wire. In just what
direction these lines of force run around the wire, we
cannot say, but to fix our ideas we may consider those
lines positive which would carry a north pole around
the wire in the direction of the hands of a watch, if
the observer sees the wire from the end, and if the
current flows from the observer. If we place a piece
of paper over such a magnet and at right angles to
such a wire, the lines of force made visible by iron
filings will now appear as somewhat distorted circles

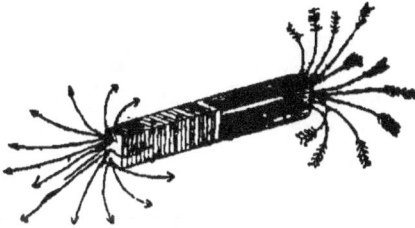

FIG. 2.

crowded densely into the poles of the magnet. Within the magnet they are not so well defined as upon
the outside of it, but as near as can be judged from
rough experiment, no lines are lost in the middle, the
same number of lines that enter at one end of the
magnet finding their way out at the other. This, then,
explains why it is that we cannot have one free magnetic pole or a magnet with unequal poles. Whence
it follows that every line of force must be a closed
line, an open magnetic line of force, being as much of

an impossibility as a continuous electric current in an open circuit. According to our conception, a positive line of force goes in the air from the north to the south pole, and in iron from south to north. If, instead of a steel or "permanent" magnet, we had

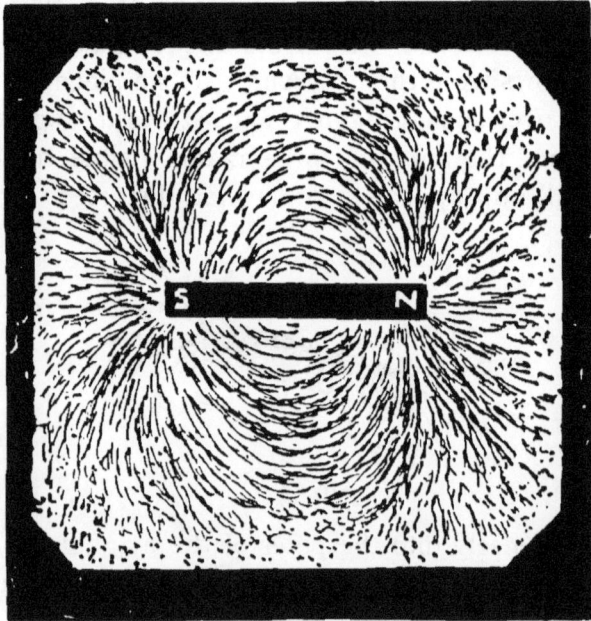

FIG. 3

experimented with a piece of soft iron, we should also find that through its presence the originally circular lines around the wire would have become somewhat distorted and crowded together at the ends of the piece of soft iron, which, like a steel magnet, will set itself at right angles to a straight wire carrying a current. The total number of lines made visible by filings would, under these circumstances, be considerably increased, even on the side

FIG. 4.

of the wire opposite to that occupied by the piece of iron; and if we should now increase the current, we would still further increase the number of lines, showing that the number of lines is dependent upon the strength of the current, and that one and the same current will yield more lines if a piece of soft iron be placed in its neighborhood, and also showing that the

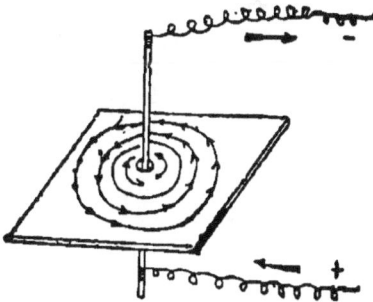

FIG. 5

lines pass more easily through iron than air. This latter circumstance leads us to the conception that every medium offers a certain *resistance* to the passage of these lines of force, and the inverse of that resist-

ance is what is called the "*magnetic permeability*" of
the medium, which, in the case of iron, is from 40 to
20,000 times as great as that of air. In other words,
if we should surround this straight wire by a cylinder
of iron, the same current would yield within that

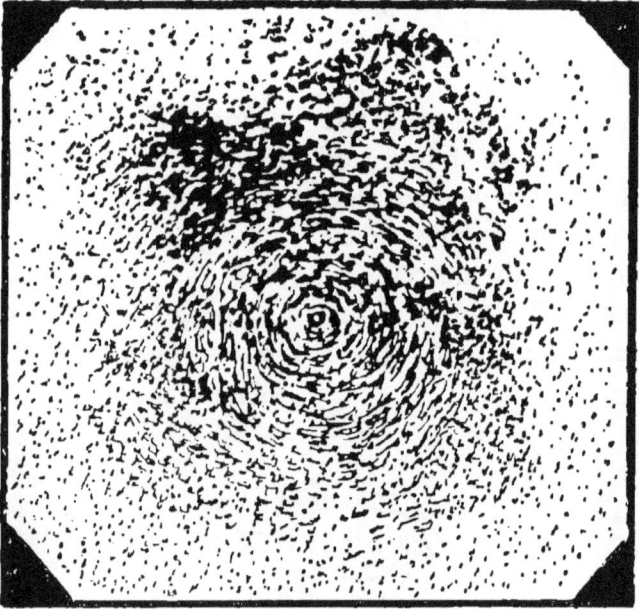

FIG. 6.

space, from 40 to 20,000 times as many lines of force
as when the space contained only air. In the same
way, any permanent magnet emits similar lines of
force, which go through the magnet from the south to
the north pole, emerging at the north pole and com-
pleting their circuit through the air by travelling

from the north to the south pole, as here shown (see Fig. 2).

These lines of force are no fiction of the imagination like the lines of latitude and longitude on the globe; they exist and can be rendered visible by iron filings sprinkled upon a card held over a magnet, or having a wire carrying a current run through its centre at right angles to the plane of its surface. These ideas are shown very clearly and diagrammatically in Figs. 1, 2, 4, 5, and 7, and more beautifully and accurately in Figs. 3, 6, and 8, the latter being reproductions from Nature, showing experiments with iron filings. We see, therefore, that the electric current may be treated as a magnetic phenomenon, and that both in the case of a magnet and in that of a wire carrying a current, a portion, at least, of the energy exists outside of the magnet, or of the wire conveying the current, and must be sought in the surrounding space. It is supposed that these magnetic "*whirls*," or lines of force, are set up by movements, pressures, and tensions in the surrounding *Æther*, as the result of the peculiar molecular vibrations constituting or accompanying the electrification of the wire. With this explanation, we are now in shape to establish the relation between a current and a magnet, and to show how the one may produce the other.

When we wind a wire carrying a current, into a spool or helix, as before described, the magnetic whirls or lines of force in the surround-

2 QQ

ing space, are no longer small circles wrapping
around each separate turn of the conducting wire,
but the lines of adjacent turns of the wire forming the helix merge into one another, and run con-
tinuously through the helix from one end to the other,
as here shown (see Fig. 7). Compare this figure with
Fig. 2, and the similarity in the arrangement of the lines
of force in a magnet is obvious. When we push a bar of
iron into the interior of this helix, the lines of force will
run through the iron and magnetise it, converting it

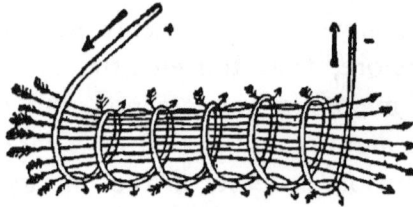

FIG. 7.

into an "*electro-magnet*," as shown in Fig. 8, where the
the lines of force, or "*magnetic field*," as it is called,
is shown as reproduced from the actual figure made
by iron filings. Having now described how an electric current, or rather a wire under the condition of
electrification, possesses for the time magnetic properties and is surrounded by lines of force, and further
shown how magnetism can be produced from electric
currents, we are naturally led to a consideration of
the converse condition of affairs, and to assume that
magnetism may produce electricity; for in Nature,

wherever we find an action, there must we also find a reaction, and we are, therefore, led to assume that if magnetism is the accompaniment of the electric current and the threading of these lines of force through the helix, as in Fig. 7, then the threading of lines of force

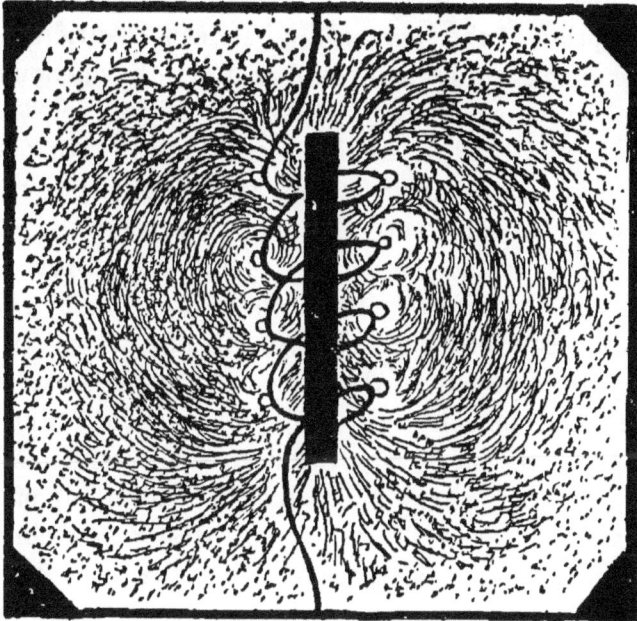

FIG. 8.

through a closed helix of wire, said lines being emitted by a magnet, should in turn produce an electric current. Such assumption is correct, and the act of introducing a magnet into the centre of such a closed helix or spool of wire will produce an electric current in the helix, since such act is accompanied by the passage of lines of force through and around said helix. This constitutes the principle upon which all magneto-electric generators are based, whether it be the Faradic coil or the electric light generator, and is known as the inductive method of generating currents.

We should now have a very correct and clear conception of the relationship between electricity and magnetism, and of the peculiar physical phenomena which accompany the manifestation of both.

THE METHODS OF MANIFESTATION of that condition known as electrification are many besides those of simple magnetism. Although electricity or electric energy cannot be felt or seen, it is nevertheless capable of manifesting itself, through the many *effects* which it can produce. These are of various kinds. They are magnetic, (which has already been considered), thermal, chemical and physiological. For instance, a current flowing through a wire will heat it, the amount of heat generated depending upon the relation between the volume or strength of current flowing, and the resistance or size of the wire; flowing over a magnetic needle it will cause the latter to turn, as we have already seen, the angle through which the needle is deflected being a measure of the volume or strength of the current flowing; flowing through liquids it decomposes them, the amount of decomposition produced in a given time being also a measure of the volume or strength of the current flowing; and lastly, flowing through the living body, or any sensitive portion of it, it produces certain peculiar sensations, as well as chemical and physical changes.

THE METHODS OF GENERATING ELECTRICITY or of setting up that condition of matter known as electrification, are numerous, and consist mainly of *induction, chemical action, thermal contact* of different metals, and *friction.* In electro-therapeutics we have to deal principally with the chemical and inductive methods.

THE TWO QUALITIES OF AN ELECTRIC CURRENT.—Electricity possesses two qualities, that of *"pressure"* and that of *"volume"* or *"strength."* With the nature of these qualities once clearly in mind, the whole science and its involved phenomena immediately become clear. For the purpose of securing a definite, tangible conception of the nature of an electric current, therefore, it may be likened to a current of water flowing through a pipe, and we can illustrate its behavior by pointing out its analogy in hydraulics. Any device for generating electricity, such as a Galvanic battery, a thermo-electric pile, or magneto-electric or dynamo-electric machine, is essentially a pump, which pumps electricity instead of water. If the discharge pipe of a hydraulic pump be carried around through a given circuit and connected into the suction side, both pump and pipe being full of water, the movement of the pump will cause the water to flow in one particular direction, producing a *"continuous current"* of water. Substitute a Galvanic battery or a continuous current electric light gener-

ator known as a *"dynamo"* for this hydraulic pump,
wire for the pipe, and electricity for the water, and
we have a conception of an electric current at once
clear and tangible as to all of its elementary phenomena. Let us now place within brackets those electrical terms which are analogous to the terms used in
hydraulics, and it may be said that a certain number
of pounds [*volts*] of pressure is required to overcome
the friction [*resistance*] of the pipe [*wire*] in order
that the water [*current*] may flow at the rate of so
many gallons per minute [*amperes*]. The larger the
pipe [*wire*] the more water [*current*] can be carried,
and the less will be the friction [*resistance or drop*];
or, *per contra*, the smaller the pipe [*wire*] the less the
quantity [*amperes*] per minute, and the greater will be
the friction [*loss or drop*]. It is at once evident, then,
that the pipe [*wire*] might be so small that the friction [*resistance*] would absorb a very large proportion
of the power of the pump [*dynamo or battery*], leaving
but little remaining for useful effect. Let us carry
this illustration of a hydraulic pump circulating or
pumping water through a continuous main pipe still
further, and assume that at various points in this
main, water motors are introduced. Motors thus
located would be said to be in *"series,"* *i. e.*, arranged
like beads on a string, one after the other (see Fig. 9).
It is apparent that when the pump is started and the
current begins to flow, the motors will also begin to
move, but the pressure developed by the pump must

of necessity be equal to the number of pounds pressure required to move each motor or to force the current through each motor, multiplied by the number of motors, in addition to the pressure required to overcome the loss in the main pipe by friction; that is to say, if the frictional resistance of the main pipe is

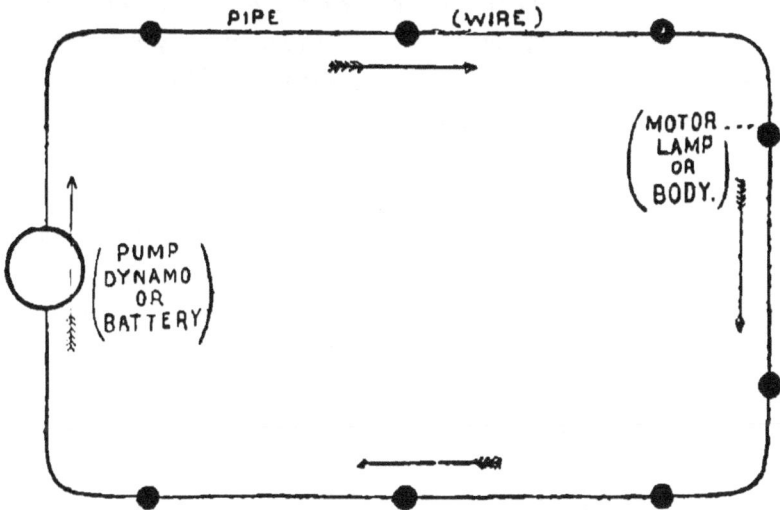

Fig. 9.

equal to ten pounds pressure, and we have a series of eight motors each requiring two pounds pressure to operate them, the gross pressure to be developed by the pump must be equal to $2 \times 8 + 10 = 26$ pounds total pressure. This condition of affairs corresponds exactly to an electric circuit through which it may be desired to send a current of electricity, and in which the water motors may be replaced either by electric motors, electric lamps, or human beings or portions of the latter, assuming them all to be in series. We see, therefore, that it is by virtue of that quality which we term pressure in hydraulics, that we are enabled to force a current of electricity through any electric circuit and overcome the latter's resistance. In electric science we use the term "*electro-motive force*" to designate the pressure or head under and by virtue of which an electric current circulates. This electromotive force is created by reason of what is called a "*difference of potential*" across the two sides of the source of the electric current, whether this be a Galvanic battery, a dynamo, or some other form of electric generator. This difference of potential corresponds to what is called a difference of level in hydraulics, or to a difference of temperature in thermodynamics. If, for instance, two vessels filled with water, be connected by means of a pipe, and one vessel be placed at a higher level than the other, a current of water will flow through the connecting pipe from the higher vessel to the lower one; and, so again, in the case of two bodies in which there exists a difference of temperature, if these be so placed that heat can pass from one to the other, heat will pass from the hotter to the colder as long as any difference of temperature exists. In like manner, a current of electricity will flow from a higher or "*positive*" potential to a lower or "*negative*" potential so long as there exists a difference of potential, and the degree

of electro-motive force which urges a current forward
will depend upon the extent of the difference of
potential which exists between the two sides of the
source of supply or between the positive and negative
poles of any form of generator. A difference of potential in electricity is, therefore, analogous with difference of pressure in gases, with difference of level in
liquids, and with difference of temperature in heat.

Now, just as we use the pound as a unit
of measure of pressure in hydraulics, so do we
use the unit known as the " *Volt* " as a unit of
measure of electro-motive force in electricity. The
standard value of the unit being that electro-
motive force which will establish or circulate a current of electricity of one " *Ampere* " through a resistance of one " *Ohm.*" * This naturally brings us to
inquire into the nature and value of these two last
mentioned units, the " *ampere* " and the " *ohm.*" The
" *ampere* " is the unit of measure of the " *volume* " or
"*strength*" of a current, which latter is directly proportional to the amount of chemical decomposition produced by the current in a given time, the standard
value of the unit being represented by that current
which will deposit 0.00111815 gramme, or 0.017253

---

* A Daniell element or battery, especially prepared for
this purpose, for example, has an electro-motive force of 1.079
volts, while a Latimer Clark's standard cell, which is very
constant on open circuit, has an electro motive force of 1,457
volts.

grain, of silver per second on one of the plates of a silver "*voltameter*," the liquid employed being a solution of silver nitrate, containing from 15 to 30 per cent. of the salt.* The term "*ohm*," on the other hand, is applied to designate the unit of measure of the "*resistance*" which any given electric circuit or part thereof offers to the passage of a current of electricity. Not only do the lines of force emitted by a current, and constituting the magnetic whirl surrounding every wire carrying a current, meet with a certain resistance in their passage through different media, as we have seen, but the current itself in passing through a circuit, no matter of what the latter may be composed, meets with a certain amount of obstruction or "*resistance*," so that by the insertion of a longer or shorter piece of wire, or of a longer or shorter column of liquid into the circuit, or of a wire of a greater or less diameter, a current can be diminished or increased in strength. Any number of amperes can be sent through any body, provided we have a sufficiently powerful generator or electromotive force to, as it were, pump or force the current through the resistance offered by the body, and provided further that the body is not fused or other-

---

* The usual strength of the currents used in telegraphing, over main lines, for example, is only from five to ten thousandths of an ampere, or five to ten milli-amperes, the prefix *milli*, indicating that the standard commercial unit, the ampere, has been divided by one thousand.

wise destroyed by the current before it has reached the required strength. The "*Ohm*," or the legal unit of this resistance, as settled by the International Electrical Congress, at their meeting held at Paris in 1884, is that resistance which is offered by a column of pure mercury 106 centimetres long, and 1 square millimetre in cross sectional area, at a temperature of 0° C. Previous to 1884 the unit of resistance used most extensively in Great Britain and elsewhere was the British Association, or "B. A." unit. The two units bear the following relationship:

1 legal ohm = 1.0112 B. A. units.
1 B. A. unit = 0.9889 legal ohm.

Since, as we have already shown, the strength or volume of a current is a function of the relationship of the electro-motive force and the resistance of the circuit through which it passes, then we should naturally surmise the existence of a law establishing the character of this relationship. Such a law was deduced by Dr. Ohm in 1827, and it has since been verified. This is now known as "*Ohm's Law*," and it may be briefly stated as follows: *The number of ampères of current flowing through any circuit is equal to the number of volts of electro-motive force divided by the number of ohms of resistance in the entire circuit.* Or, algebraically expressed:

$$\text{Current} = \frac{\text{Electro-motive force (in Volts)}}{\text{Resistance (in Ohms)}}$$

Or,

$$C. = \frac{E.}{R.}$$

It must be borne in mind that the resistance here referred to is the whole resistance of the entire circuit, *i. e.*, the resistance not only of the circuit external to the source of supply, or generator, but also of the source of the current, or generator itself. For instance, if a number of cells of a battery are used, and the circuit be made up of a number of different parts through all of which the current must flow, we have to take into account not only the combined electromotive forces of the cells, but the combined resistance of all the parts of the circuit, both within the cells and without. That is, the current may flow from the zinc plate of the first cell, through the liquid to the copper (or carbon) plate, then through a connecting wire or screw to the next cell, through its liquid, through the other connecting wires and liquids of the remainder of the cells, then through a wire to a Galvanometer, for instance, then through the wire coils of the Galvanometer, then perhaps through the whole or a part of the tissues of a human being, and finally through a return wire to the zinc pole of the first battery. In this case there are a number of separate electro-motive forces, all tending to produce a flow, and a number of different resistances, each impeding the flow and adding to the total resistance. If in such a case we knew the separate values of all the different electro-motive forces, and all the different resistances, we could calculate what the current would be, since it would have the value—

$$C = \frac{E' + E'' + E''' + \ldots \ldots}{R' + R'' + R'' + \ldots \ldots}$$

$$\text{Or,} \quad C. = \frac{\text{Total Electro-motive Force.}}{\text{Total Resistance.}}$$

Where E represents the electro-motive force of each of the cells, and R the resistance of each portion of the circuit. Without a thorough comprehension of this law, which forms the basis of the whole science of electricity, it is impossible to deal intelligently with the science. Once having a thorough and clear conception of it, however, we are enabled to explain all of the varied phenomena. Knowing any two of the factors in the equation representing Ohm's law,

$$C = \frac{E}{R},$$

that is, knowing the resistance and the electro-motive force, we may, by calculation, determine the third factor, or the current; or knowing the current and the resistance in any given case, we may, by calculation, determine the electro-motive force; or knowing the electro-motive force, we may, by calculation, determine the resistance; since the *current* equals the electro-motive force divided by the resistance,

$$C = \frac{E}{R}.$$

while the *electro-motive force* equals the current multiplied by the resistence ($E = C \times R$), and the *resistance* equals the electro-motive force divided by the current,

$$R = \frac{E}{C}.$$

THE SO-CALLED VARIETIES OF ELECTRICITY differ from each other only by reason of the preponderance of one or the other of these two qualities of pressure and volume. All the different so-called varieties will produce the same effects under similar conditions and relationship of resistance and pressure. If properly insulated, so-called static electricity, for instance, will flow and produce a current which will magnetize iron; and if it has sufficient volume it will decompose a liquid or heat a wire. If two static machines be placed upon opposite sides of a room, with their respective poles connected together by means of well insulated wires, and one machine be revolved by some power, the machine on the opposite side of the room will immediately begin to revolve, just as in the case of the transmission of power by means of two dynamo-electric machines, the one acting as generator and the other as a motor. The current here flows from one machine to the other. The current from an induction coil is also in every respect similar to a Galvanic current, except that it is intermittent and its direction constantly reversed, now flowing in one direction and then in the opposite direction. If, however, the current be "*commutated*" and caused to flow always in one particular direction, it is capable of producing the same chemical decompositions and magnetic effects that are produced by a Galvanic current. It has the same two qualities of pressure and volume, and the volume is a function of the relationship of electro-motive force and resistance, just as in the case of the Galvanic current. Hence it should be apparent that the effects produced by a Faradic battery, which is nothing but an induction coil, will depend entirely upon the volume or strength of current employed, which is likewise subject to Ohm's law, $C = \frac{E}{R}$—. Hence the strength of any Faradic current employed for medical purposes will depend solely upon the electro-motive force generated by the Faradic battery employed, divided by the total resistence of the circuit through which the coil is operating. This resistance will be made up of the internal resistance of the coils of wire of which the Faradic battery is composed, the resistance of the connecting wires, and also the resistance of the subject or of the tissues through which the physician may be passing the current. There is nothing mysterious about it. There is nothing that cannot be dealt with mathematically. Hence all the confusion of ideas and contradictory conceptions which prevail in the minds of most electrotherapeutists regarding the character of the current from different Faradic batteries and from coils composed of different sizes of wire, is all unnecessary. Here, as with the straight Galvanic current, it is

purely a question of the strength or volume of current employed, and this is measurable in terms of the same unit which we employ with the Galvanic current, namely, the milli-ampere. With these principles in mind it will be seen that it does not necessarily follow that a coarse wire coil will give us the greatest strength or volume of current through all resistances.*
It is customary, however, for the electro-therapeutist to make the mistake of always looking to the coarse wire coil for a great volume of current. The conditions of the relationship of resistance to the electromotive force may frequently cause us to obtain a current of greater volume from a fine wire coil than could be derived from the coarse wire coil when operating through the same resistance, and *vice versa*. The electro-motive force or pressure of the current derived from an induction coil or Faradic battery is principally dependent upon two things:

1. The strength or volume of the current flowing through the primary coil, any variation of which will cause a corresponding variation in the electromotive force of the current generated in the secondary coil.

2. The number of turns or convolutions of wire composing the secondary coil —*no matter what the*

* Any more than it necessarily follows that a Galvanic battery arranged for "quantity" will always give the greatest strength or volume of current, irrespective of the resistance of the circuit through which it has to circulate.

size of the wire may be, an increased number of turns increasing the electro-motive force generated in the secondary coil. Of course, in order to get a great number of turns of wire in a given space, fine wire must be used, but the *size* of the wire has nothing to do with the character of the current generated. The fine wire secondary coil, while it will give a much higher electro-motive force, will also have a much higher internal resistance, and the strength or volume of current it will cause to flow will depend entirely upon the relationship of this electro-motive force to the total resistance of the entire circuit, *i. e.,* the internal resistance of the coil and the resistance of the parts or tissues through which the current is flowing. How absurd it is, then, for the physician to undertake to say that a current of great volume has been used simply because the large wire coil was employed. The failure to comprehend the nature of the problem involved here has given rise to many contradictory reports as to the comparative therapeutic effects of the coarse and fine wire secondary coils of a Faradic battery. For instance, Dr. G. Betton Massey, whom we all know well, writing to me under date of April 19th, 1890, says: "In the physics of the subject, I trust you will throw some light on the coarse wire nonsense in our induction currents as produced in the batteries we use. Although I am a disbeliever in a secondary coarse wire, and in my book speak slightingly of the primary in-

3 99

duction current, I have recently noted undoubted evidence of high muscle-contracting power with the latter current, when the resistence of the circuit was minimized by very large poles." Could there be better proof of the truth of the principles just stated, or a better example of the prevailing chaotic and contradictory notions concerning this subject? Of course Dr. Massey noted better muscle-contracting power from the coarse wire coil in this particular instance! Why? Simply because the external resistance was low, as he states, and therefore he could get a great volume of current with a low electro-motive force such as is generated by the coil of fewer turns or "large wire coil." When we approach or withdraw the secondary coil to or away from the primary one, we respectively increase or decrease the electromotive force of the current generated in the secondary coil, by doing what is *equivalent* to adding to or taking away from the number of turns of wire composing the secondary coil. What we really do, however, is to place more or less turns of wire within the "*field of force*" of the primary coil, the internal resistance of the Faradic battery remaining the same all the while.

To sum up then, the whole question is purely one of *strength* of current, which must be *measured* in order to have any definite idea concerning its value, and thus be able to reproduce the same conditions a second time. It is to be regretted that we have never heretofore had an instrument for measuring the Faradic current. In the present work, however, the author will have the pleasure of presenting to the profession for the first time such an instrument, which has been christened the *"Faradometer."* This instrument will be illustrated and described later on.

THE QUANTITY OF ELECTRICITY which will flow by a given point or into a given body in any given time, will naturally depend upon the *volume* or *strength* of the current flowing, just as in the case of a stream of water flowing into a reservoir, where the quantity of water emptied into the reservoir in any given time depends entirely upon the volume or size of the flowing stream. Hence the *quantity* of electricity which flows through or into a body is equivalent to the product of the *volume* or *strength* of current (number of amperes) multiplied into the *time* the current is flowing. There is then a distinction between the *quantity* of electricity used and the *volume* or *strength* of current employed.

THE COULOMB is the established unit of measurement of the *quantity* of current, and its standard value is equivalent to that quantity of electricity which will flow through or into a body when one ampere of strength or volume flows for one second of time. A two ampere current flowing through or into a body for one second of time, or, on the other hand, a one ampere current flowing for two seconds, would represent a *quantity* of current equal to two *coulombs*. The physician who wishes to use electricity as a therapeutic agent with precision, and who desires to make and record accurate observations and thus secure uniform results, will have as much occasion to use this unit of quantity as he will have to use the unit of strength of current, for the one is of but little value without the other.

ELECTRICAL UNITS OF MEASUREMENT are no simple fiction of the imagination. They bear a fixed and determined value and relation to the units of measurement of all other forms of energy, such as heat, light and dynamic force, and are convertable one into the other.

The principle of the correlation and conservation of energy, announced for the first time by Helmholtz, controls all problems in the measurement of force, and plays an important part in the application of the system of absolute units established by the British Association, and to the consideration of which the Congress, at the recent electrical exhibition in Paris, devoted much attention, with the object of establishing the system on an International basis. The term *"energy"* is applicable to all physical manifestations— mechanical work, the production of heat, light, etc Conservation results from the important fact that energy expended is always to be re-found integrally in some other expression of work—calories, chemical action, and so forth. It is evident, then, that the measure of the various forms of energy ought to be based upon a system of units intimately dependent one on the other and convertible one into the other, but the establishment of such a system is the work of time; several centuries, for instance, were required to create such a revolution as would sweep away existing prejudice and permit of the establishment and logical classification for the first time of a series of geometrical units for the measurement of space and weight.

Nearly a century ago, a National convention in France marked a great era in progress by the creation of the metric system, which takes as a basis of measures of length, area, volume, weight and mass, the metre, which is equal to the forty-millionth part of the length of the terrestrial meridian. This was the first step in the direction of establishing a universal and inter-convertable system of absolute units. The next step was taken by the British Association when it based the system of electrical measurements on the units of length, mass, and time, adopting for each the *centimetre*, the *gramme*, and the *second*, thus forming what is universally known as the centimetre, gramme, second, or C. G. S. system of units.

This C. G. S. unit of force is *that force which acting on a gramme of matter for a second, generates a velocity of one centimetre a second.* This unit is named the *Dyne*, and is equivalent to the $\frac{1}{980}$ of the weight of a gramme of matter at any part of the earth's surface, or, in other words, 980 *dynes* is about equal to the weight of one gramme of matter. This is then the C. G. S. unit of *force.* It was necessary to establish another fundamental unit to represent the *energy* or *work done* by any *force.* The C. G. S. unit of *work* or *energy* represents the work done in exerting a force of one *dyne* over the distance of a centimetre. This unit with the above value has been named the *Erg.* These two fundamental C. G. S. units of *force* and *energy*, then, constitute the foundation for all electrical units and measurements, and we shall now trace their relationship to the units of all other kinds of energy, whether manifested in the form of mechanical work, heat or otherwise, and show how they are convertible one into the other. The most generally employed expression of mechanical power, for instance, is the *horse-power.* It was introduced by that great thermo-dynamic engineer, Watt, and is equal to 33,000 foot-pounds per minute, which is equivalent to 76 kilo-grammetres, since the foot pound is about one-seventh of a kilo-grammetre. In France the horse-power is called the "*cheval vapeur*," and is equivalent to 75 kilogrammetres. Now, the kilo-grammetre, by the C. G. S. system is equal to 98,100,000 ergs, and the "*cheval vapeur*" is, therefore, equal to 7,357,500,-000 ergs. To avoid the use of numbers altogether too large to be convenient, such as the *absolute* C. G. S. units involve, the Congress adopted a series of *commercial* units, consisting of the *Volt*, the *Ampere* and the *Ohm*. The *volt* is equal to one hundred millions of absolute C. G. S. units of electromotive force, or ten to the eighth power (expressed 10^8); the *ohm* is equal to one thousand millions of absolute C. G. S. units, or ten to the ninth power (expressed 10^9). By this time the reader should have a very clear idea of the definite and immutable character of these electrical units, and of their fixed relations to the units employed in dealing with all other forms of energy; and it should be impossible for him to commit the error embodied in the following statement made by a recent author of considerable note, whose work is before us, and from which we quote: "*A milli-ampere is resistance offered by a human body to a current of electricity which is generated by the Daniell's element.*" The author in question might with equal correctness have said that a gallon is equal to a gas-pipe, or a mule's hoof to a boy's aching stomach.

Electric power is the product of the pressure or electro-motive force of a current in *volts* multiplied into the strength or volume expressed in *amperes*.

THE WATT is the name of the unit employed to express this product of the *volt-ampere*, and it is equivalent to one seven hundred and forty-sixth ($\frac{1}{746}$) of a horse-power; that is to say, seven hundred and forty-six (746) watts equal one horse-power; hence the volts of pressure multiplied by the amperes of strength of any current, the product being divided by

746, will give the power of the current in "*horse-power.*" Whence the formula :—

$$H. P. = \frac{\text{Volt} \times \text{Amperes}}{746} \text{ or } \frac{E \times C}{746} = H. P.$$

With this further information properly assimilated, the reader should be able to appreciate the fallacious character of the statement contained in a paper recently read before the American Medical Association, which was to the effect that the author of the paper had frequently used a TEN AMPERE (not *milli*-ampere) current in the treatment of uterine fibroids, the current being passed through the abdominal walls to an electrode placed in the tumor per vagina. The very lowest possible resistance offered by the tissues of the body in a case of this kind, no matter how large the abdominal electrode, would at least be not less than sixty ohms. Now by Ohm's law ($E = C \times R$) we may know that the electro-motive force which must of necessity have been employed in order to force this ten ampere current through the sixty ohms of resistance offered by these tissues, was ten multiplied by sixty, which gives six hundred volts. And now, since the current in amperes multiplied by the electro-motive force in volts gives us the number of watts of power employed, we have ten multiplied by six hundred as the number of watts. This product, which equals six thousand, when divided by the constant 746, gives us the horse-power of energy which must have been expended for five minutes of time upon any such patient. Now just for one moment imagine the whole of EIGHT HORSE-POWER of *any* form of energy—whether it be in the nature of the mechanical power of a shaking machine, heat, chemical action, or electricity, being expended upon a human being and that person still living! We do not recite this circumstance to the disparagement of the writer of the paper in question, for we are well acquainted with him, and know him to be thoroughly conscientious and honest, but rather for the purpose of giving a single instance in illustration of the practical value to an electrologist of a thorough knowledge of the science of electricity, and the necessity for the exercise of great care in the selection and use of electrical measuring instruments.

Having now discussed in a general manner the nature of that condition known as electrification, and the laws governing the circulation of electrical currents, we shall immediately proceed to describe the *technique* and apparatus necessary for the scientific application of this form of energy to the medical and surgical uses of the physician and surgeon, leaving the many and varied electrical and electro-magnetic laws and phenomena with which we shall be obliged to come in contact, for consideration in connection with the practical application of the knowledge, feeling that in this way these laws and phenomena will be the better impressed upon the mind of the reader.

CHAPTER III.

THE APPLICATION OF ELECTRICITY IN MEDI-
CINE AND SURGERY.

THE NECESSARY PARAPHERNALIA for the scientific application of electricity in medicine and surgery should comprise:

1.   A generator or source of supply from which to derive the "direct," or Galvanic current.

2.   A current controller of proper design and construction for regulating the strength of the current.

3.   A milli-ammeter of proper design and construction for accurately measuring the STRENGTH *or* VOLUME *of the Galvanic current.*

4.   A coulombmeter for measuring the QUANTITY *of the direct or Galvanic current employed in any case.*

5.   A volt-meter for occasionally determining the total electro-motive force of the generator or source of supply, or of individual cells of a battery, and for use in measuring differences of potential of various parts of the circuit, as a means of checking up other measurements and measuring instruments.

6.   A direct reading ohm-meter, or a box of standard resistances, for measuring the resistance of different parts of the body at different times, and for making tests of the resistance of various other parts of the circuit.

7. An adjustable rheotome for interrupting either the Galvanic or Faradic currents any required number of times per minute.

8. An induction coil of suitable range for generoting alternating currents of varying electro-motive forces.

9. A Faradometer for measuring the strength of the Faradic currents being employed at any time or in any given case.

10. An electro-static induction machine for generating currents of small strength but extremely high electro-motive forces. ·

11. Various other accessories, such as switches, pole-changer, connecting-cords or rheophores, suitable variety of electrodes, electrode-handles, etc.

The contention is not made that all of these things are absolutely necessary to the simple use of electricity. One or all may be dispensed with, just as many practitioners entirely neglect the use of electricity in medicine. But what we do contend is, that the degree of success which any practitioner will meet with in the practice of medicine, will largely depend upon the extent to which he utilizes this most important form of energy called electricity, and that its scientific administration with a view to advancement and the establishment of it upon an exact rather than an empirical basis, demands the use of all of this apparatus, just as much as does the study of bacteriology require the use of at least a good microscope.

THE GENERATORS OR SOURCES OF ELECTRICAL

SUPPLY from which the medical practitioner may derive electricity for medical and surgical uses are:

    a.   Central Power Stations.

    b.   Individual Dynamo-Electric Generators.

    c.   Galvanic Cells.

(*a*) CENTRAL POWER STATIONS OR ELECTRIC
LIGHT MAINS, have recently been resorted to as a
source of such supply. Although among the first, if
not *the* first, to employ currents from such stations,
the author unhesitatingly pronounces in favor of the
primary battery for general electro-therapeutic work,
and recommends the use of currents from electric
mains only for the operation of electric motors and
the charging of so-called storage batteries to be used
in heating cautery electrodes, maintaining electric
lamps, and operating electric motors outside of the
office, and inside the office in case of a "shut-down"
at the central station.

Unless specially designed for the purpose, and
provided with an "armature" (the revolving coil of a
generator) having a great number of sections and
run at a positively uniform speed, the ordinary
dynamo-electric generator such as we find in these
stations does not furnish an electric current suitable
for general therapeutic purposes. The current is
physically the same as that derived from a Galvanic
cell, and will produce the same chemical and physiological effects, but it is too fluctuating for use as a
therapeutic agent. It is impossible, without produc-

ing unbearable pain, to use the same strength of current in any given case that could be given were a battery current employed. Its use involves too many chances for painful and dangerous shocks.

The use of storage or secondary batteries for general electro-therapeutic work is precluded by reason of the high electro-motive force which is generally required for therapeutic purposes. The number of ordinary storage cells which would be required to establish this necessary electro-motive force would be so great as to make a battery of enormous weight and great cost. Storage cells, again, are apt to give a great amount of trouble, require very careful handling, and are subject to instantaneous discharges and even complete destruction through some slight inadvertency on the part of the operator. Hence their use should be resorted to only for those purposes which occasionally demand a very strong current, such as Galvano-cautery, etc.

(*b*) INDIVIDUAL DYNAMO-ELECTRIC GENERATORS with a great number of armature sections have been designed, and in a few instances employed by physicians, to give currents for general therapeutic, illuminating and cautery purposes. The first machine of this kind was designed and used by the author in 1887. The machine was made according to the author's design by the Excelsior Electric Light Company of New York, and is capable of giving either of two currents upon a simple changing of some plugs

upon the top of the machine; one of the two currents
has an electro-motive force of four volts and a possible strength of forty amperes, and is suitable for all
cautery and illuminating purposes; while the other
current has an electro-motive force of forty volts and
a possible strength of four amperes, and is suitable
for most therapeutic purposes. Machines similar to
this are now regularly manufactured and upon the
market as a staple article. The best machine of this
kind is illustrated in Fig. 10.

This machine generates two distinct currents at
the same time, the armature being a compound one
with two separate and distinct windings, each of
which are respectively connected with two separate
and distinct commutators, so that both currents may
be utilized for different kinds of work at the same
time, or either one separately. One of these currents
is identical with that generated by a cautery battery
of cells having elements with large surface, while the
other current is similar to that generated by a large
number of cells arranged in "series;" the two are
designated respectively the "*cautery*" and the "*electrolytic*" current. The maximum electro-motive force
of the former is five volts, with a possible strength of
sixty amperes; while the maximum electro-motive
force of the latter is two hundred volts, with a possible current of one ampere. Either current may be
varied from zero to full power by a simple rotation of
the "*rocker-arm*" or "*brush-holder*" of the corre-

sponding side. An automatic safety device in the
nature of a fuse is supposed to prevent the current

FIG. 10.—THE MEDICAL DYNAMO.

H E—Handles for rotating brushes on the electrolytic side.
H C—Handles for rotating brushes on the cautery side.
S S—Thumb screws for setting brushes.
A—Armature.
K K—Brushes.
C P—Positive post of cautery current.
C—Negative post of cautery current.
T—Commutator.
X—Safety fuse.
E P—Positive post of electrolytic side.
E—Negative post of electrolytic side.
C and *E P.*—Plugs for increasing or diminishing the current.
P L—Driving pulley.

from reaching a strength that would be calculated to harm a patient, destroy a cautery electrode, or injure the machine itself. This machine will generate a current that is steady and reliable, and that is equal to that produced by the largest cautery battery made for office work, while at the same time producing another current with sufficient electro-motive force to operate a 16-candle power incandescent lamp, or do any work that can be accomplished by a battery of 150 Lelanché cells. This machine should be run at a uniform speed of 1,800 revolutions per minute, by either an electric motor, a gas engine, a water motor, or a steam engine of one-horse power, preferably the former, if electric mains are at hand. If one has a large special practice, and is *permanently* located near electric mains that are kept energized the whole twenty-four hours, and if the care and manipulation of electric machinery is quite well understood by the would-be user, then the medical dynamo may be recommended as a very desirable source of electric supply. Our advice to the general practitioner, however, is to let the medical dynamo religiously alone.

(*c*) GALVANIC CELLS OR BATTERIES are pieces of apparatus or devices for converting *chemical* energy into *electrical* energy. The production of electrical energy is, strictly speaking, only the conversion of one form of energy into another form of energy, or one mode of motion into another mode of motion. All forms of apparatus which produce electrical en-

ergy at the expense of chemical action, are called
" Galvanic Cells," and an aggregation of such cells is
spoken of as a *"battery"* of cells or as *"batteries."*
While there are many varieties of these cells upon
the market, there are, nevertheless, but few which
may be said to meet the requirements of the medical
man. It is a chimerical fancy to suppose that any
one form or kind of battery will meet all of these requirements. One might with equal reason and success seek in one form of apparatus a source of electrical supply which would be equally suited to the
varied purposes of *electro-chemical deposition, gilding,
silvering,* the production of *arc* and *incandescent lights,*
the operation of *converters* or *induction coils,* the discharge of *torpedoes,* the propulsion of *aëreal ships,* and
the operation of *telegraph lines, bells* and *telephones;*
each of which requires a special form or kind of generator adapted to meet the peculiar exigencies of the
work to be performed. Just as in the case of these
several industries, the medical man will in his office
require one form of battery for charging *"secondary"*
or storage batteries, another form for *cautery,* motive
power, and illuminating purposes, and another for
neurological and gynæcological work; and different
forms of all of these for work outside of the office.
The writer has had a wide experience with many
hundreds of cells, involving almost all types, and covering a period of something like fifteen years. In
addition to this practical experience, the author has

4 QQ

recently examined and made careful comparative electrical tests of all the latest leading forms of such types of cells as would be likely to meet the requirements of the medical profession, with the object of being able to give a definite and reliable opinion in the present work.

Galvanic cells may be divided into two principal classes: *Open Circuit Cells* and *Closed Circuit Cells*. Those which come under the first class are the only ones which are suitable for therapeutic purposes. What is commercially known as the *Leclanché cell* is the typical representative of this class, and of the many forms of it which we find upon the market, the two best are the "*Gonda*" and the "*Axo*" cells manufactured by the "Leclanché Battery Co.," of New York (see Figs. 11 and 12).

FIG. 11.—"GONDA" BATTERY COMPLETE.

FIG. 12.—"AXO" BATTERY COMPLETE.

The Theoretical Conditions of a Perfect Battery are:

(1) *A high electro-motive force.*

(2) *A low and constant internal resistance.*

(3) *A constant electro-motive force irrespective of the current produced by the cell.*

(4) *A consumption of inexpensive materials.*

(5) *A lack of consumption of all material when no current is being produced, that is, when the circuit is not closed.*

(6) *A ready means of occasionally examining its condition and working, and of adding fresh materials when required.*

It should always be borne in mind—

That the electro-motive force of a Galvanic Cell is independent of its *size*, a cell no larger than a thimble giving as high an electro-motive force as one the size of a barrel.

That the character of the elements employed dedermines the electro-motive force of the cell, all cells having similar elements giving *practically* the same electro-motive force.

That one cell will give practically the same strength of current on a short-circuit as one hundred similar cells arranged in series.

That no greater strength of current can be gotten out of one hundred cells arranged in series than one cell will give on a short-circuit.

That one hundred cells arranged in series will,

however, force the same strength of current through one hundred ohms of external resistance that one cell will create through one ohm.

That the strength of current which any cell will give is dependent on its internal resistance, which latter is entirely governed by the extent of surface of the elements, their proximity to each other, and the character of the solution employed, an acid solution offering less resistance than an alkaline one.

The Leclanché cell proper is composed of a glass jar, a cylindrical rod of zinc as a positive element, and a porous cell containing about equal quantities of manganese dioxide and gas carbon as a negative element. The glass jar is half-filled with a saturated aqueous solution of ammonium chloride, or *sal ammoniac*. There is no chemical action till the circuit is closed. Its electro-motive force varies from 1.2 to 1.5 volts, and the internal resistance varies from .5 to 5 ohms. It contains no acids and no poisonous substances, gives off no acid vapors and no odor; the materials used are cheap and easily replaced, and resist the most intense cold; so that, as a type, it more nearly fulfills the above conditions than any other form of cell.

The *"Axo"* cell is the best pattern of this regular standard type. A modification of the Leclanché, has recently been devised, in which the porous cell is replaced by a conglomerate block composed of a mixture of forty parts of manganese dioxide, fifty-five of carbon, and five of gum-lac, the whole being subjected to a pressure of three hundred atmospheres at 100° C. The "*Gonda*" cell is the best and latest form of this modified Leclanché cell. This cell has an electro-motive force of 1.5 volts, which is unusually high, with as low an internal resistance as any other cell of its kind. It "*polarizes*," that is, its internal resistance increases very slowly under action, and it "*depolarizes*," that is, resumes its original resistance, very rapidly under rest. The positive pole outside of the battery, it must be borne in mind, comes from the negative or manganese and carbon element, while the negative pole comes from the positive or zinc element. These open circuit cells require but little attention, and one charge of the solution will last from six months to several years, according to the amount of work imposed upon them. Even with daily use upon a number of patients, the zinc element will last at least one year. The time will surely come, although it may not be for quite a while, when the manganese-carbon element must also be renewed. There has, however, more recently still been introduced a form of cell which does away with this porous cup of manganese and carbon, substituting a simple element of gas carbon having a very large surface. Such elements with proper care are practically everlasting. The very best pattern of this form of open circuit battery is represented by what is commercially known as the *Double Cylinder* "*Law*" battery manufactured by the Law Telephone Co., of New York, and shown in Fig. 13.

The negative element (positive pole) is formed of a double cylinder of gas carbon, which furnishes a

FIG. 13.—THE DOUBLE CYLINDER "LAW" BATTERY.

very large surface that reduces the internal resistance and enables a ready escape of the polarizing gases, thus requiring no depolarizing agent such as manganese. This negative element of the *Law Battery* is practically everlasting. In all porous-cup batteries

which require a periodical renewal of the negative
element, the first cost of the cell is insignificant as
compared with the subsequent cost of these renewals.
Both the carbon and zinc (negative and positive)
elements of this battery are permanently attached to
a composition cover closely resembling hard-rubber
(see Fig. 14), which is so constructed that by a slight
turn it locks down upon the jar tightly with an inter-
vening soft rubber ring, thus effectually sealing the
jar and preventing a rapid evaporation of the solution and a crawling or creeping out of the ammonia
salt. The absence of such a device in all other forms
of this type of battery constitutes a serious defect, and
the effort to accomplish the same end by coating the
top of the jar with a greasy substance like paraffin, is
both ineffectual and dirty. The jar of this battery is
more compact and contains more fluid for the cubic
space occupied than that of any other. Both connections are alike, and of the most compact and best
form. It is elegant in appearance, and its mechanical
construction is of the highest order. Our own measurements show that the electro-motive force averages
1.35 volts, although the makers claim 1.5. The current which it will give upon a " *short-circuit*," that is,
through an inappreciable external resistance, varies
from 1 to 2½ amperes. Although the electro-motive
force of this cell is slightly below that of the " *Gonda*,"
taking into consideration its compactness, its perfect
mechanical construction, its air-tight seal, its cleanli-

ness, its indestructible negative element, and the beauty and symmetry of its outline, we would recommend it above all others as the best form of stationary battery for medical purposes. In all open circuit batteries both elements remain in the solution all

FIG. 14—COVER AND THE TWO ELEMENTS OF THE LAW BATTERY.

the time, and there is no consumption of any of the materials except when the circuit is closed and the current flowing, and even then the consumption of materials is strictly in proportion to the strength of the current being generated, or, in other words, inversely proportional to the resistance through which the batteries are working. This peculiar feature admits of the use of all of the cells or the full electromotive force of the battery at once, introducing a large artificial resistance to cut down the strength of the current; this resistance may then be gradually withdrawn until the required strength of current is attained. Such a variable artificial resistance for regulating the strength of the current is known as a current *"controller."*

THE CURRENT CONTROLLER is indispensable in all Galvanic work, and frequently very useful in Faradic work. Its office is to turn off and on the current, steadily and gradually, and thus prevent disagreeable and injurious shocks in all cases where these are to be avoided. It is nothing more nor less than a *" rheostat"* which interposes a certain initial artificial resistance, which may be very *steadily* and *gradually* withdrawn and reinserted under the manipulation of the operator, thus correspondingly increasing and decreasing the strength of the current, a graduation that is impossible with the old *" cell-selectors,"* *" wire-coil rheostats,"* and *"primitive water rheostats"* that it supplants. Its principal advantages over the old wire-coil rheostats, and cell-selectors are:

1. Its simplicity; avoiding all the complicated wiring incidental to the others.

2. Its greater certainty of preventing shocks.

3. Its distribution of the wear equally amongst all the cells of the battery, which is a most desirable feature.

4. Its saving of the necessary exhaustion of each cell as it is short-circuited by the selector-crank in turning on and off cells by means of the current-selector.

5. Its need of but two terminal wires, which renders it possible to connect a stationary battery with a movable table holding the apparatus.

The "Bailey Current Controller," Fig. 15, made by the Law Telephone Co. of New York, is probably the best controller at present upon the market, although Dr. G. Betton Massey, of Philadelphia, in his recent admirable work on "Electricity in the Diseases of Women," describes one of his own design, made by Otto Fleming of same place, which he claims is superior to it; and from his description we can certainly recognize several points of advantage. We only hope that the uncertain and perishable character of the pencil mark used as a resistance in the Massey controller, will not prove an offset to these advantages. The Bailey controller consists of four carbon plates which are mounted so as to be readily immersed in a jar of water by means of a rack and pinion. Two of these plates are connected together metallically to form a pair, and the pair is connected with one of the binding-posts of the apparatus. The remaining two plates are also connected together to form another pair, and these are again connected with the other binding-post. The two pairs of plates are insulated one from the other,

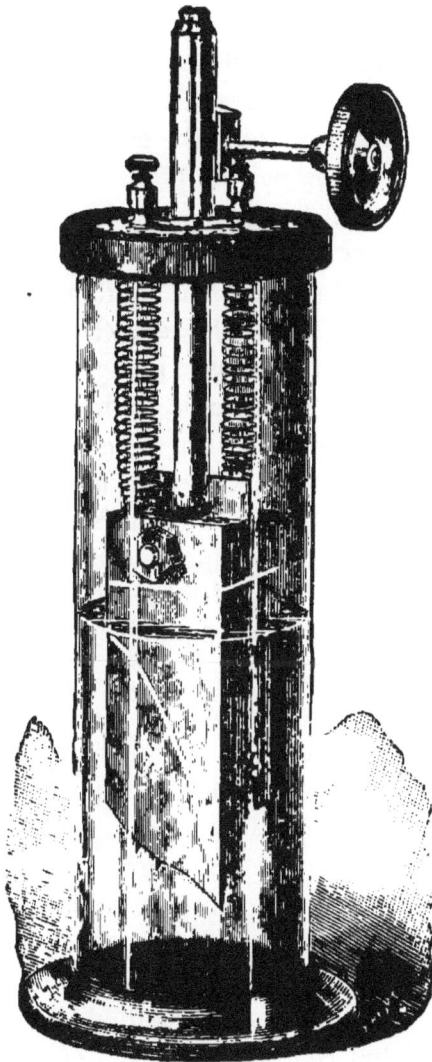

FIG. 15.—BAILEY CURRENT CONTROLLER.

and the current must pass from one pair of carbon plates to the other through the water. When entirely out of the water, no current is supposed to flow. Of course the resistance which the current meets with, in its passage from one pair of plates to the other, will depend upon the extent of surface of the plates brought in contact with the water, and this is varied by raising and lowering the carbon plates in the water. As found upon the market at the present time, this controller, however, possesses the following serious defects: The metallic connections within the jar soon corrode, thus creating a very high resistance, finally crumbling to pieces and breaking the connection entirely; by capillary attraction and the condensation of moisture, a film of water is soon formed upon the insulating strips between the pairs of carbon plates, which forms a conducting bridge for the continuous passage of the current even before the carbon plates touch the surface of the body of water, thus causing a disagreeable shock in delicate electrizations, and frequently even completely destroying the function of the controller; owing to an improper shape of the carbon plate, the gradations of the current's strength are not sufficiently gradual and regular; when the carbon plates are completely immersed in the water there still remains from ten to thirty ohms resistance in the circuit, which should be entirely removed. All but the last of these defects may be removed in the following manner: Unscrew the carbon plates from the suspending rod and rack, and after subjecting them to a dry heat equal to the temperature of melted paraffin, immerse the upper end of the carbon plates, connections and all, in the melted paraffin, down to a point indicated by the line Z in Fig. 15, keeping them in the hot paraffin for several minutes, then gradually allow the paraffin to cool and remove the plates. Before doing this, however, take the plates apart, and with a fine key-hole saw cut the lower ends in the manner indicated by the line X in Fig. 15, so that the corners formed by the opposite pairs of plates will form a paraboloid. Remove the sponge tips. Bore a fine hole in the lower end of each carbon plate, and insert therein a piece of the lead from a pencil, about one-half inch in length.

The plates, when viewed from the side, will now present the appearance shown in Fig. 16. When the plates are so treated, and made into this shape, the whole condition of affairs is changed, and the Bailey controller becomes a very valuable and sensitive piece of apparatus. The connections will not corrode, and when the plates are now slowly immersed, we not only very gradually increase the surface of the plate exposed to contact with the water, but we also gradually decrease the length of the path that the current has to travel in going from one pair of plates to the other, which, at the start, is from A to B, or from one pencil point to the other. It will be seen from the figure that this distance is very gradually decreased, as well as the exposed surface increased as the plates are immersed to a greater depth. With the Bailey controller as modified by the author, the current may be turned on *progressively* and *regularly*, without a jump

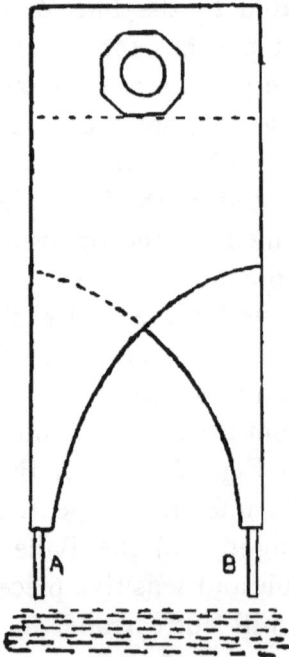

FIG. 16.—THE AUTHOR'S MODIFICATION OF THE BAILEY CONTROLLER.

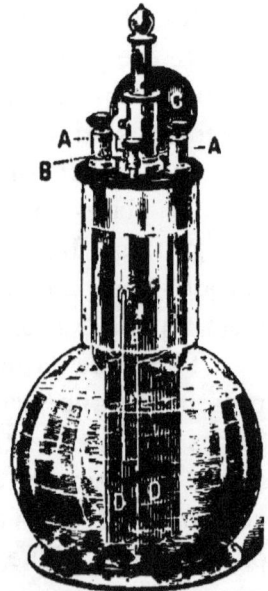

FIG. 17.—THE McINTOSH CURRENT CONTROLLER.

or an oscillation, at the rate of only one one-hundredth of one milli-ampere for each increment of current strength. There then remains but one defect, which consists in the inability to finally remove all of the resistance without a sudden jump.

Another form of controller has been placed upon the market by the McIntosh Battery Co., of Chicago [see. Fig. 17], but we do not think well of it; besides, its cost is excessive. Between two small sheets of platinum (D, D) suspended in water, with suitable attachment (A) for one pole of the battery, is suspended a third piece of platinum (E) that can be lowered or elevated by means of a rack and pinion (B, C).

THE MILLI-AMMETER is the most important and delicate instrument that we have to deal with. A trained physicist, who is acquainted with all of the influencing conditions, can, in his laboratory, determine the strength of any electric current by means of an ordinary Galvanometer, because such a Galvanometer —of a particular shape and size, and with a definite magnetic needle, acted on by a definite controlling force produced, for instance, by the earth's magnetism, or by some fixed permanent magnet—has a perfectly definite law connecting the magnitude of the deflection with the strength of the current producing it. But such an instrument cannot be carried around everywhere, and everybody cannot use it. The scale is only divided into the degrees of a circle, whereas the strength of current is proportional to the tangent of the angle of deflection, so that in any case of measurement with such an instrument, the tangent of the angle of deflection must be sought for in a table of tangents, which necessitates a complicated calculation involving as factors the component of the earth's magnetism at the place, and the "*constant*" of the Galvanometer. To avoid the necessity of such a complicated process, and enable anybody to read directly from an instrument the strength of current flowing through it in terms of the unit of current strength (the ampere), ordinary Galvanometers have, for some time, been made, in which the ordinary degree divisions have been replaced by divisions the lengths of

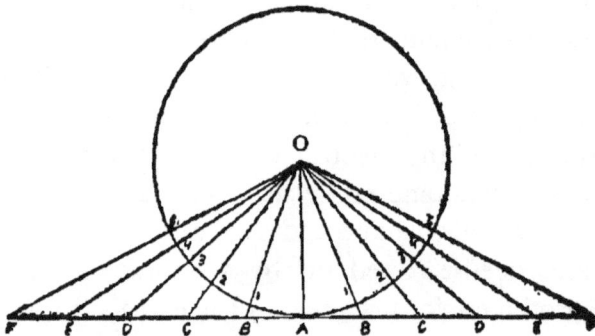

FIG. 18.

which become smaller and smaller as we depart from the zero or undeflected position of the needle, in such a way that the number of divisions in any arc is proportional to the tangent of the angle corresponding with that arc. The accompanying diagram (Fig. 18) will make this clear.

The lengths A B, B C, C D, etc., along the line A F, which is tangent to the circle at the point A, are all made equal to one another; hence, if from the

centre (O) of the circle, straight lines, O A, O B, O C,
etc., be drawn, cutting the circumference of the circle
at the points A, 1, 2, 3, etc., the numbers 1, 2, 3, 4,
etc., will be respectively proportional to the tangents
of the angles A O 1, A O 2, A O 3, etc. The spaces
between the lines drawn from the center of the circle
(O) to the various points A, B, C, D, etc., on the line
A F, are equi-distant, but at the points where they
cut the circle at 1, 2, 3, etc., they rapidly approach
each other as they are drawn further from the vertical
line O A. Suppose now that we wind a Galvanometer with a coil of wire of such size or of such a
number of turns that a current of ten milli-amperes
will deflect the needle from its zero or resting point
(A) to point 1 on the circle, then a current of twice
this strength or twenty milli-amperes should deflect
the needle to the point 2, because the line O C cuts
the circle at this point, and C is the same distance
from B that B is from A. In like manner, thirty
milli-amperes would deflect the needle to 3, and forty
to 4, etc. The limit of this proportionality in most
instruments is reached at from 50° to 60°, and beyond
this point the needle moves over such a small distance for large increments of current that measurements in this part of the scale are of but little value.
This fact greatly limits the range of such instruments.
This same supposed instrument may be rendered
more sensitive by using a greater number of turns of
a finer wire for the windings, so that only one milli-

ampere will deflect the needle to the point 1 on the circle, when five milli-amperes will deflect the needle to the point 5, because the point F is just five times the distance from A that the point B is. In like manner the instrument may be made less sensitive

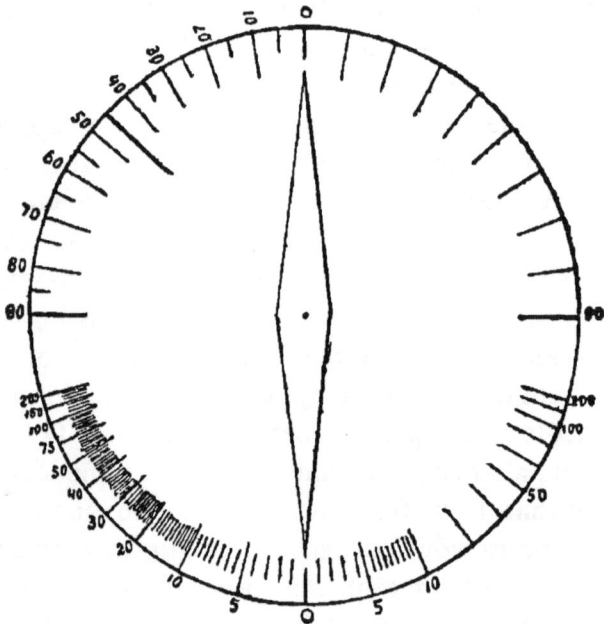

FIG. 19.

than in the first case by using fewer turns of a coarser wire than was first used, so that a current of one hundred milli-amperes will only cause a deflection of the needle to the point 1 on the circle, in which case a deflection of the needle to the point 5, will indicate five hundred milli-amperes or one-half an ampere. It will be observed that in all three of these cases the law of tangent proportionality remains the same, in each case a movement from A to 2 indicating twice as much current as the movement from A to 1, although in each case the actual amount of current causing this movement may differ, in the one case the strength of current being two, in the other twenty, and in the third case two hundred milli-amperes.

The difference between an ordinary Galvanometer whose scale or dial is divided into simple de-degrees, and one which has its scale or dial divided so as to read the current strength directly, is clearly shown in Fig. 19, where the dial in its upper half is divided into degrees, and in its lower half into milli-amperes. The principle that the angle of deflection does not increase proportionally to the strength of current, is illustrated by the fact that whilst a current of 30 milli-amperes deflects the needle to about 45°, a current of 150 milli-amperes is required to deflect the needle to 70°.

Such a direct reading instrument as we have just described, operating against the controlling influence of the earth's magnetism, upon the principle simply of an ordinary Galvanometer, is known as a *"calibrated"* Galvanometer. Such instruments have been made and placed upon the market by various instrument makers at home and abroad, notably, Gaiffe, Mc-Intosh, Waite and Bartlett, and called *"Direct Reading Ammeters"* and *"Direct Reading Milli-Ammeters."*

As a matter of fact, however, such instruments are not, in the strict sense of the word, "*Direct Reading Ammeters.*" As we have already stated, the Galvanometer employs the earth's magnetism, as the controlling force against which the strength of any electric current is balanced, hence this force must be constant for all time and places, in order to have an instrument calibrated in one place and at one time, read correctly at all other places and at all other times. Now, as a matter of fact, the force of the earth's magnetism is not constant for all years, nor for all parts of the globe. The subjoined table will give an idea of its variations as regards both time and place:

PLACE.	YEAR.		
	1870.	1875.	1880.
Paris...................	1 94	1.96	1 98
London.................	1 78	1.80	1.82
Leipzig................	1.86	1.88	1 90
Darmstadt..............	1.91	1 93	1.95
Edinburgh......	1 62	1 64	1.66
Zurich.................	2.00	2 02	2 04
Dublin	1.67	1.69	1.71
Turin	2.07	2 09	2.11
Vienna.................	2.05	2.07	2 09
Königsberg ....,	1.79	1 81	1.83

Since the angle of deflection produced by a given Galvanometer is evidently proportional to the directive or controlling influence of the magnetic force acting on the needle, it is obvious that the indication of any Galvanometer calibrated in London, for instance, would be excessive when used in Turin or Vienna, in the proportion of 1.82 to 2.11 and 2.09 respectively, or deficient when used in Edinburgh or Dublin. Hence correction must be made for different places and different times when using such instruments, and they cannot, therefore, be called *direct* reading instruments in the strict sense of the word. Again, the presence of any small amount of inductive material, such as iron, steel, or artificial magnets, in the neighborhood of such instruments, causes an error in their reading. It has, therefore, been found necessary to substitute some other controlling force than the earth's magnetism against which to balance the electric currents in direct reading measuring instruments. Such instruments are called "*Ammeters*," when they are designed to measure currents over one ampere in strength; and milli-ammeters, when designed to measure fractions of an ampere.

The controlling forces employed in the various types of such instruments have been: *Permanent Magnets;* the *Force of Gravity;* and the *Elastic Force of Springs*. In all of these different types, the law of action has varied. In some the deflection has been directly proportional to the strength of current, and only in very few cases has the law of tangent proportionality been employed, because of the practical objection to it that the *divisions are unequal* and therefore confusing to the reader. Of recent years, some of the brightest intellects amongst our physicists and electrical engineers, both at home and abroad, amongst the number being Sir William Thomson, Marcel Deprez, Siemens, Cunynghame, Ayrton and Perry, Schuckert, Crompton and Kapp, Edward Weston, Elu Thomson, and others, have been diligently at work endeavoring to design a type of ammeter which should be delicate and accurate and meet all the practical requirements of a commercial instrument. The problem has been one of the most difficult of solution ever submitted to the physicist. And yet the makers of shoddy, junk electro-medical apparatus, have flooded the market with worthless instruments that have been palmed off upon the credulity of physicians as reliable "*milli-ammeters.*" An ammeter must be an instrument that indicates in terms of the *absolute universal standard unit* at all times and in all ordinary places, or it is nothing more than a "*Galvanoscope,*" which simply indicates the *presence* of a current. To be a commercially practical instrument it must possess the following qualifications:

It must have a wide range:

It must be permanently frictionless in the practical sense:

It must be absolutely "*dead-beat*;" that is, the needle must not vibrate or oscillate under variations of the current, but move promptly up to the point marking the increase or decrease in the strength of current, and there *remain* without vibration or oscillation, just as though it were being moved by a very fine threaded screw:

It must not be affected materially by variations in temperature:

Its readings should not vary materially 'with age:

It should read correctly for all places:

It should be *delicate* or *sensitive*, and at the same time have a *low* resistance, which is a very difficult combination:

It should be absolutely direct reading:

Its law should be, that the angle of deflection varies *directly* with the strength of current, so that the *divisions* upon the scale or dial may be *equal* in length for equal increments of current strength, in all parts of the scale:

It should be portable:

It should not require levelling or other adjustment to bring the needle to the zero mark:

Its zero mark should not require to be placed in the line of the magnetic meridian of the earth:

Its needle should be deflected over a considerable and easily appreciable distance for slight increments of current:

It should have a double scale, and an accurate means of adapting it to suit different ranges of current strength:

It should not be seriously affected by the presence of neighboring pieces of iron or magnets:

Its needle should always return promptly and exactly to the zero mark on shutting off the current, without oscillation or vibration, and there remain, without having to tap or jar the instrument:

It should, above all things, be absolutely accurate in graduation, and indicate correctly in all parts of the scale, in terms of the absolute and universally recognized standard unit of current strength:

It should not be necessary to tap the instrument in order to start the needle:

Its needle should have a very small *moment of inertia*, and should move in a very powerful controlling field:

It should not be affected by neighboring wires conveying currents:

It should *not* possess any *magnifying gearing*, because these introduce friction and add to the moment of inertia of the moving parts, and so diminish the "*dead-beat*" character of an instrument.

Of all the attempts that have been made by foreign physicists to design an instrument that would meet the above requirements or qualifications of an ideally perfect instrument, the honor of the accomplishment has fallen to our countrymen, Prof. M. M. Garver and Mr. Edward Weston, of Newark, N. J. This instrument is manufactured by the Weston Electrical Instrument Co., of Newark, New Jersey, and is known as the *Weston Ammeter*. It has been commended and adopted by nearly all of the leading electrical engineers both at home and abroad. It was designed with a view to meet the constantly increasing demand, in the commercial use of electricity, for a portable instrument of greater range, accuracy, and reliability, than had heretofore been obtainable. All previous forms of ammeters were subject to marked variations and uncertainties, resulting from high temperature errors and general defects in electrical and mechanical design and construction. The electrical, magnetic, and mechanical features of this instrument are such as to eliminate all such variable elements, and ensure permanence and reliability coupled with simplicity, extreme accuracy, and portability.

Being familiar with this instrument in its application to the commercial uses of electricity, the author of this book communicated with the Weston Electrical Instrument Co. with a view to having this Company place upon the market a modification of their commercial ammeter that would indicate *milliamperes* and meet the *special* requirements of the medical profession. This resulted in an acceptance of our suggestions, and the construction of an instrument with modifications in accordance with our design. This instrument is illustrated in Fig. 20.

This is the first absolutely correct, direct-reading, "*dead-beat*," milli-ammeter that has ever been placed upon the market. As compared with it, all previous so-called milli-ampere meters may, without exaggeration, be said to be mere pieces of junk-shop apparatus. The use of this instrument will gladden the heart of any experienced electrologist, or even the most scientific and exacting physicist.

It is *absolutely* " dead-beat," the needle moving *promptly* up to the proper mark without oscillation, and remaining there.

FIG. 20.—THE WESTON MILLI-AMMETER.

Made in accordance with the Author's suggestions and designs, to meet the requirements of the medical profession.

The needle always returns promptly to the zero mark.

It does not require tapping, levelling, placing in the line of the magnetic meridian, or other adjustment. It is provided with two scales, a *black* and a *red* one. The upper or *black* scale reads to five hundred milli-amperes or one-half of an ampere, each division representing five milli-amperes.

The lower or *red* scale reads to ten milli-amperes or one-hundredth of one ampere, each division representing one-tenth of a milli-ampere.

This uniform decimal arrangement of both scales greatly facilitates the readings and makes it impossible to make an error in the act of reading; for instance, in the lower scale we read one, one and one-tenth, one and two-tenths, etc., two, two and one-tenth, two and two-tenths, and so on up to ten; while in the upper scale, we read five, ten, fifteen, twenty, twenty-five and so on up to five hundred. The divisions are far apart or of great length, and easy to read, the needle moving over about $\frac{1}{16}$ of an inch to indicate only one-tenth of a milli-ampere, and over about eight inches in indicating only ten milli-amperes. By this scale arrangement the needle is generally moving in about the centre of the scale when doing electro-medical work with either the upper or lower scale, the centre of the scale representing the 250 and the 5 milli-ampere mark in the upper and lower readings respectively. The instrument is provided with three binding posts, one neutral in color for the positive wire, one black, and marked 500 for connection with the negative wire when using the upper or black scale, and one red post for connection with the negative wire when using the lower or red scale, thus making an error regarding the scale employed impossible.

All the divisions of the scale are equal in length, thus preventing confusion.

Although extremely sensitive, the resistance for both scales is very low, being for the upper scale only .19 (nineteen-hundredths) of a legal ohm, and for the lower scale only 10.43 (ten and forty three-hundredths) legal ohms.

For all ordinary purposes no temperature correction is needed. The actual difference for 35° above or 35° below 70° F., is only about one per cent., being subtractive for an increase, and additive for a decrease in temperature. All these instruments are separately and individually standardized at the Weston Laboratory by a competent expert, and certified to by Mr. Edward Weston, one of the most competent and widely known electrical engineers of this country. It is a very compact instrument, and is put up in a handsome cherry box with a leather handle, which greatly adds to its portability. The author has examined and made careful tests of all the leading forms of so-called milli-ammeters now upon the market, and can without hesitancy declare this to be the only one deserving of the confidence of the profession, or which even deserves to be called a " *Direct-Reading, Dead-Beat Milli-ammeter.*" It is also so constructed that it may be turned into a *volt-meter* by simply throwing an artificial resistance into circuit with it, thus making it a combined volt and milli-ammeter. A neat little resistance box accompanying the instrument contains two resistance coils, one bringing the resistance of the instrument up to

10,000 ohms, and the other up to 1,000 ohms. When the 10,000 ohms coil is thrown into circuit, the red scale being used, the instrument will read to 100 volts, each division representing one volt, and when the 1,000 ohm coil is thrown in, it will read volts as well as milli-amperes. Here then, we have a very unique electrical instrument, measuring from a tenth

Fig. 21.—The Barrett Milli-Ammeter.

of a milli-ampere to five hundred milli-amperes, and from a tenth of a volt to one hundred volts, with an accuracy seldom found even in the best physical laboratories.

The only other instrument that is deserving of mention in this category is the Barrett Milli-ammeter (see Fig. 21). This is a cheaper instrument, and compared with some of the other so-called milli-ampere-meters upon the market, an admirable one; but it is neither accurate, reliable, nor "dead-beat," and compared with the Weston instrument it is almost worthless. The purchase of a cheap milli-ammeter will be found to be exceedingly poor economy in the end. It is impossible to build an instrument which will be both cheap and reliable. They require a certain amount of hand work, and the time of an expert in adjusting and callibrating them, and such labor is necessarily expensive. An instrument that has no hand work about it, and that does not require individual adjusting and callibrating, is necessarily a worthless one, and it would be better to do without any, than to rely upon such a one. We are just upon the eve of establishing electro-therapeutics upon an exact and scientific basis, in lieu of the empiricism of the past, and a false step at this period, through the use of unreliable measuring instruments, would be disastrous.

The current or milli-ampere meter should always be placed in "*series*" with the remainder of the circuit, as shown in the diagramatic illustration of the circuit connections for the Author's electro-therapeutic cabinet (Fig. 29, Vol. II). Figure 9 also illustrates a "*series*" arrangement for electrical devices. The office of the milli-ampere meter is to at all times reveal to the operator in terms of the standard unit of current strength the exact volume or strength of current that is being dealt with and which may be flowing through a patient or any portion thereof, thus enabling the operator to work intelligently and to make an exact record of the strength of the current employed in any given case, for future reference, or for purposes of publication and comparison. It is absolutely impossible to judge of the strength of current used in any case by the number of Galvanic cells employed, since the current is always a function of the relationship of the resistance of the entire circuit to the electro-motive force of the battery, and this is constantly changing and rarely the same. Ten cells may one day have twice the electro-motive force that they had upon the preceeding day, or ten cells may even one minute have twice the electro-motive force they had the preceeding minute. The internal resistance of battery cells is also constantly changing, so that a cell or a battery of cells may one moment have twice the internal resistance that they had a moment before. The degree of polarization of a battery has much to do in determining its internal re sistance and electro-motive force, and this is constantly changing, particularly in the case or open-circuit batteries. Again, the resistance of the external circuit is also subject to constant changes. The patient or subject being operated upon constitutes a part of this external resistance, and we rarely find two different subjects presenting the same resistance between corresponding points, or even the same subject presenting the same resistance between similar points upon two different occasions. The size, character and moisture of the electrodes employed, has much to do in determining the resistance; the connections at the "*binding-posts*" and elsewhere throughout the circuit, also have much to do in fixing the resistance, a poor connection at a switch or binding-post often introducing as much or more resistance than the remainder of the circuit, including the resistance of the patient.

In the construction of current meters for medical purposes, the milli-ampere. which is the one-thousandth part of one ampere, has been fixed upon as the standard unit of calibration, because this unit and its multiples correspond to the strength of the currents used in medical applications of electricity. For example, the resistance of those parts of the human body included between medium sized electrodes applied to the spots commonly selected, averages from one to four thousand ohms, and a current of one milli-ampere would be yielded by one to four Daniell's cells, and this is *about* the weakest Galvanic current ever used therapeutically or diagnostically.

In dealing with drugs we use and speak of solutions of a certain strength, and we use solutions of various strengths for the production of different effects ; the strength of any solution is, however, no indication whatever of the *dose* or *quantity* of the drug which may have been administered. In like manner in dealing with electricity we use and speak of currents of a certain strength, and we use currents of various strengths for the production of different effects; here also, the strength of any current is, however, no indication whatever of the *dose* or *quantity* of electricity which may have been administered. Just as in the case of drugs, the dose of electricity is a question of the *quantity* administered at any one time, and this is not alone dependent upon the strength of current employed. To two different subjects we may wish to administer the same dose or quantity of electricity, and yet employ currents of widely different strengths in the two cases, using in the one case, for example, a current of ten milli-amperes, and in the other one, a current of one hundred milli-amperes. In dealing with solutions of drugs we employ graduates for measuring out the quantity to be employed.

THE COULOMBMETER is an instrument capable of performing an analogous service in the case of electricity. This instrument, and not the milli-ampere-meter, as some writers have stated, measures the *dose* or *quantity* of electricity employed.

An instrument of this character has recently been constructed and placed upon the market by M. Gaiffe, of Paris, see Fig. 22. It is made of two glass tubes, one within the other, which are designed to be filled with water, and corks placed in their respective mouths. The current passes into the instrument

6 QQ

through one 'of. the platium wires (F), then through
the water across to the other platium wire (F')

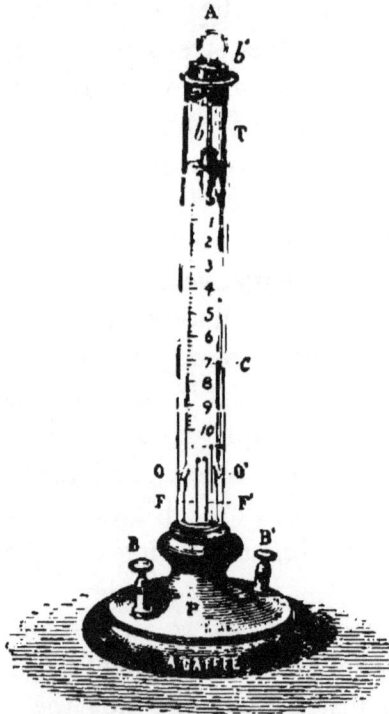

Fig. 22.—Direct Reading Coulombmeter.

T outer tube, C inner tube, both fixed to a stand, P.
—b and b' corks closing the upper extremities of the tubes.
b can be lifted out of the inner tube by pulling the wire A
which passes through the cork b'.

O, O' apertures in the inner tube; F, F' platinum wires
projecting into the inner tube and connected with the binding screws B, B' to which the wires of the external circuit are
attached. The inner tube is graduated so as to read directly
in coulombs.

and so out, in its passage decomposing the water into oxygen and hydrogen, which ascend in the inner tube and collect at the top, thus depressing the water in the inner tube, the level of which marks the number of coulombs passed through the instrument. After the instrument has been used, and gas has formed within the inner tube, expelling a portion of the water, equilibrium may be re-established and the water from the outer tube caused to flow back into the inner one, by simply lifting the inner cork by means of the wire (A), when the instrument will be ready for another measurement. A small quantity of water must occasionally be poured into the outer tube, so as to insure the refilling of the inner tube when its cork is raised.

This instrument reads directly up to ten coulombs. In ordinary neurological work the doses range from ½ to 6 coulombs, while in gynecological work they range from 4 to 200. This instrument does not read high enough, and possesses too high a resistance, for use in gynecological work, without the addition of a " *shunt*" circuit or by-path around the instrument, of comparatively low resistance, such as will multiply the reading say twenty times, which will enable it to read up to 200 coulombs, and at the same time decrease the resistance, so as not to consume an undue amount of the available electro-motive force. A switch may be introduced to throw in and out this "*shunt*"-*circuit.* With such an arrangement we meas-

ure a certain per centage only of the total current
employed, say $\frac{1}{20}$th. Such an arrangement is shown
in the diagrammatic illustration of the connections for
the authors' cabinet, Fig. 29, Vol. II.

By the use of this instrument, then, in conjunction
with a milli-ampere meter, we may observe and record
the fact that at certain specified intervals we administered to a patient for some particular affection, in a
particular manner, so many coulombs of a constant
or continuous current of electricity of an *"intensity"*
of so many milli-amperes;* do not however confound
this word *"intensity"* with the term *"tension,"* as is so
frequently done.†

The term *"intensity"* is used in France to indicate the strength or volume of a current of electricity,
and is synonymous with the term *"current"* as used in

---

* Just as when dealing with drugs we speak of having
administered a certain *"quantity"*–say so many drops or tea-
spoonsful–of a solution of a certain *"strength"*–say five, ten
or twenty per cent. or even a saturated solution of a particular drug, at specified intervals. Here, then, we have presented a means of dealing with electricity as a remedial
agent in a scientific and precise, rather than in the empirical
manner of the past. The possibilities and the developments
which such a method promises, must be at once apparent to
any accurate observer.

† For instance, Bartholow in his work on "Medical
Electricity," page 37, says: "By *intensity* of a current is
meant its power of overcoming resistance." Others make
the same error.

this country. In France, for example, they write *"intensity"* or $I. = \frac{E}{R}$, while in this country we write *"current"* or $C = \frac{E}{R}$.

The term *"tension,"* on the other hand, is employed to designate the pressure or head under and by virtue of which a current of electricity flows, and is synonymous with the term *"electro-motive force."* A clear comprehension of this distinction will elucidate much of the existing confusion to be found in current writings.

THE VOLT-METER is nothing more or less than an ammeter constructed of extremely fine wire, so as to have a very high resistance as compared with other parts of the circuit. It gives by direct reading the value in volts of the difference of potential across terminals or between any two points of a circuit between which it may be inserted as a *"shunt."* It should possess the same qualifications specified as essential in a good ammeter. In the author's adaptation the Weston ammeter, the coil of wire for the red or ten milli-ampere scale is the same as that used in the Weston volt-meter, so that by the introduction of a suitable additional resistance outside, this instrument maybe converted into a volt-meter, thus presenting two instruments in one. A neat little resistance box has been made for the writer at his instance by the Weston Co, containing two separate coils of wire which bring the resistance of the instrument up to one and ten thousand ohms respectively (See 12, Fig. 28, Vol. II).

When the one thousand ohm coil is thrown into circuit with the coil for the red scale, the instrument reads in volts as well as milli-amperes, reading up to ten volts, each division representing a tenth of a volt. Throwing the ten thousand ohm coil into circuit with the coil for the red scale, enables the instrument to read up to one hundred volts, each division representing one volt. This makes a very unique instrument—a combined volt and ammeter, measuring from a tenth to 500 milli-amperes, and from a tenth to 100 volts, with an accuracy seldom found even in the best physical laboratories.

The volt-meter is of service in determining the electro-motive force of individual cells and of the entire battery or source of supply; in determining whether our storage battery needs re-charging; in determining the proper number of cells to use in connection with certain electric illuminating instruments; and in determining the difference of potential across any two points at which we may be applying a current to a patient, from which we may arrive at the resistance of the parts, and thus check up measurements which we may probably also have made with the *"resistance-box."* To one who wishes to follow in beaten paths and rely wholly upon the assertions of others, the volt meter is of but little use. To one, however, who wishes to be an original investigator, proving and correcting the assertions of others, mapping out new paths, checking up his own investigations, and inventing new appliances and methods of treatment, a volt-meter is indispensible.

THE OHM-METER OR BOX OF STANDARD RESISTANCES, is as essential for the intelligent dealing with electricity and the manipulation of electrical apparatus, as are the level and compass respectively to the builder and mariner. Electrical resistances are generally measured by balancing a known against an unknown resistance, through the agency of what is termed the *Wheatstone bridge*, but since this is not suitable for medical purposes, the writer will not burden this small book with a description of it. Direct reading *"ohm-meters"* that can be placed directly in the circuit, partly in *"series"* and partly in *"shunt,"* and from which the resistance of the circuit may be at any time read off directly in ohms, have been proposed by Werner Siemens and Fleming Jenkins, and constructed by Prof. Ayrton, nevertheless we have never seen one, and are not aware that any have as yet reached this country. It is possible these may prove most suitable for the purposes of the electrotherapeutist. The best method with which we are at present acquainted is, however, what is known as the *Substitution* method, which consists in the act of withdrawing from the circuit the unknown resistance to be measured, and gradually substituting a known resistance, until we bring the current back to the same strength that it had in the first instance, this being indicated by the reading of the ammeter included in the circuit. The amount of known resistance introduced will then be equal to the unknown resistance which has been withdrawn. This *"known resistance"* should be made up of a suitable variety of coils of fine German-silver wire, made by some reliable physical apparatus maker, and having values varying from $\frac{1}{10}$ to 10,000 ohms, arranged in such a manner as to enable the rapid selection of any intermediate number of ohms for introduction into the circuit.* The author has designed such a *standard resistance-box* in connection with a suitable device and arrangement for conveniently and rapidly throwing into circuit any required amount of resistance between $\frac{1}{10}$ and 10,000 ohms, and this constitutes a part of the cabinet shown in Fig. 28, Vol. II., being depicted at 6 in the illustration. The required resistance is introduced by a proper manipulation of the two arms and the plugs. Standing as shown in the illustration, it is in circuit ready for use, but all resistance is *"short-circuited"* (cut out) by the arms. To introduce any required amount of resistance we have only to move the arms to the proper position, and properly insert one or more of the plugs into the proper holes between the buttons.

In the practice of both electro-diagnosis and

---

* What are ordinarily called *"rheostats"* will in no manner serve this purpose, since they are not designed for it, not being made properly, and the coils not possessing standard values.

electro-therapy, we frequently have occasion to determine the comparative resistance of the tissues and muscles upon the two sides of the body, and to compare the resistance of these at one time with their resistance upon another occasion. But this is not its only value; we *constantly* have occasion to use a method of measuring electrical resistance in checking-up and proving our observations. The absence of, and failure to use, such devices and methods, on the part of would-be practitioners of electro-therapy in the past, has constituted the one potent instrumentality for the introduction of the hundreds—yes thousands—of fallacies which we find in electrotherapeutic literature.

THE ADJUSTABLE RHEATOME is a device for interrupting the circuit any required number of times per minute. The Du Bois-Reymond induction coil, presently to be described, has attached to it a slow interrupter for varying the rapidity of the inductions, but it provides no means of knowing the number of interruptions it makes per minute. In treating the various forms of paralysis and in making electrical diagnoses, it is very desirable to have a ready means of interrupting either the Galvanic or Faradic circuit any required number of times per minute, so that we may make a record of the results obtained under these varying conditions. We may wish, for instance, to record the fact that at the beginning of the treatment of a particular case of paralysis, certain muscles responded to say sixty impressions or impulses per minute of a twenty milli-ampere Galvanic current, and after a month of treatment these same muscles responded to one hundred impulses per minute of a ten milli-ampere Galvanic current, showing a *"modal"* change; or we may wish to make a similar record regarding the action of the Faradic current. Again, we often wish to send a series of rhythmical shocks with varying periods of intermission through portions of the central nervous system, with a view to producing a profound impression.

The best instrument of this kind now upon the market is manufactured by the McIntosh Battery Co., of Chicago, and is depicted at 13 in Fig. 28, Vol. II. The arrangement of circuits should be such that this rheatome will be permanently connected with both the Galvanic and Faradic circuits without interfering one with the other.

THE FARADIC OR INDUCTION COIL should have a wide range, possess an attachment for both rapid and slow automatic interruptions, a hand-switch for opening and closing the primary or battery circuit, a switch for throwing into circuit either the primary or secondary coil, and some suitable means of varying the electro-motive force of the generated current. An admirable instrument which fulfills these requirements is manufactured by the Law Telephone Co., of New York, see Fig. 23. This instrument has an extra large coil, which gives a wide range of electro-motive force. It is particularly adapted for office work. The current is varied by moving the secondary coil towards and away from the primary one, through the agency of a rack and pinion device. When the secondary coil is entirely removed from over the primary one, the current from the secondary coil is at its weakest, and that from the primary coil is at its strongest; the reverse being the case when the sec-

Fig. 23.—Du Bois-Reymond Coil. Manufactured by the Law Telephone Co., N. Y. This coil has a switch for opening and closing the primary circuit, which is not shown in the cut.

ondary coil completely covers the primary one. A double scale aranged parallel with the axis of the coils is supposed to enable the operator to reproduce a given current or one corresponding in strength to one which may have been employed upon some previous occasion, but it is misleading. Although we may place the coils in exactly the same position with relation to each other upon two different occasions, we do not thereby necessarily secure the same current, for this is governed by many other factors, for instance, the strength of the battery current energizing the primary coil, the character of the interrupter contacts, the number of interruptions per minute, and the resistance of the external circuit, including the patient, electrophores or conducting-cords, electrodes, switches, etc. The only way of knowing anything definite about the strength of the current from an induction coil, is through the agency of a current meter specially designed for the measurement of the weak alternating currents such as are generated by medical induction coils. This has not heretofore been done, but it is nevertheless possible and has been accomplished by the author. In the instrument here illustrated, the primary coil is made up of 150 feet of No. 18 wire, and the secondary coil of 3,400 feet of No. 36 wire. The secondary coil gives a mean current upon a short circuit of from $\frac{1}{20}$ to one milli-ampere. Dr. G. Betton Massey of Philadelphia, who has written one of the most practical and accurate special works on electricity in medicine that has been written up to this date, makes the following statement upon page 86 of his book.

Experimental proof of the inappreciable volume of Faradic currents. (*Experiment* 15.) Place a milli-ampere meter in circuit with the secondary coil by inclu.ling it directly between the poles of the battery, and turn on the full strength of the current; the meter will not show even the fraction of a milli-ampere.

This conclusion is erroneous, because the premises are wrong. It is assumed that the ordinary milli-ammeter is adapted to the measurement of the "*alternating*" current, such as is generated by an induction coil. Whereas, as a matter of fact, the ordinary milli-ammeter is only designed for the measurement of continous currents flowing in one particular direction. If an alternating current from a commercial induction coil of sufficient strength to operate ten 100 volt 16 candle power incandescent lamps, had been caused to flow through this measuring instrument (assuming the wire capable of carrying it),the result would have been the same—the needle or indicator would have stood still—and yet about seven thousand milli-amperes would have in this case been flowing through the instrument. The volume of the current from a medical induction coil, is of course, small, but it is nevertheless appreciable, and can be measured in terms of the same unit that we employ in measuring the Galvanic current.

The greater the volume of a current *flowing through* a muscle, the more violent is its contraction; but the mere fact of a *coarse-wire* coil being used, does not indicate that a current of great volume has been sent *through the muscle*. The vol-

ume of the current sent *through the muscle* is purely a
question of the relationship of the resistance of the
parts to the electro-motive force of the current. In
one case, a *fine-wire* coil will give the more violent
contractions, and a *coarse-wire* coil less violent contractions, while the reverse will be true in another
case. A number of coils of different sized wire are,
then, only of advantage in *increasing the range* of our
battery, or of our available electro-motive force.
A current of small volume and high electro-motive
force, such as will *ordinarily* be derived from a coil
made up of fine wire, will be less serviceable as a
muscle contractor, and more serviceable as a stimu-
lant and for the relief of pain, not because the wire is
fine or because the current generated from a *fine-*
wire is necessarily different from that generated from
a *coarse-*wire, but because there are a greater number
of convolutions or turns of the fine-wire to be cut by
lines of force, which increase the electro-motive force,
while the long coil of fine-wire offers so great a resistance that the resulting current is cut down to a very
small volume. It should, therefore, be apparent that
the same results that would be derived from a number
of *small* coils made of different-sized wire, may be
secured by the use of one *large* coil made of medium
sized wire, with means for introducing a variable resistance into our external circuit, such as is afforded
by the current controller already described. The induction coil illustrated in Fig. 23, in conjunction with

the current controller, which should be arranged to be introduced into circuit with it at will, and with the induced current to be derived from the static machine to be described in Vol. II, will furnish as great a range of currents as could be desired for all medical purposes. The slow and rapid interruptors of this coil are admirable pieces of mechanism, and in every other respect it is a superior piece of apparatus, both physically and mechanically.

While there is *physically* no difference between the poles of a Faradic coil, since the positive and negative poles are constantly alternating between the two binding-posts, the changes occuring with each alternation or change in the direction of the current, which take place with each make and break of the primary circuit, there is, nevertheless, a great difference *physiologically*. This paradoxical statement will be understood when it is explained that only one of the induced currents (that which is made on breaking the primary or battery circuit) has any appreciable nerve or muscle contracting power, and this, of course, always flows in one particular direction. This arises from the fact that the phenomenon known to electricians as *self-induction*,* operates to cut down the electro-motive force of that induced current which is generated on making the primary circuit, to such an extent that it has not the necessary

*See articles 908, 910 and 923, pages 834, 836 and 859 respectively in Ganot's Physics, 1883.

pressure to force itself through the tissues of the human body. Let us now see how this explanation suits the phenomena as we find them: Attach two electrodes to the secondary wire of a Faradic coil; now place the electrode attached to the pole marked positive upon some indifferent part of the body— either the calf of the leg or the thigh—and the other or negative electrode over the motor point for the *tibialis anticus*. Now throw into circuit the slow interrupter, close the primary circuit, and then gradually increase the electro-motive force of the secondary circuit by moving the secondary coil further and further over the primary one, up to the point where the *tibialis anticus* just begins to contract; now reverse the poles by means of a pole-changer, without removing the electrodes,—this makes the electrode over the motor point a positive one; no contraction of the muscle will now be observed. Now make an abrasion of some portion of the body and place the negative electrode upon it, the positive electrode resting upon any other part of the body; turn on the rapid interrupter, and gradually increase the secondary current, as before, up to the point where considerable pain and irritation is felt at the point of abrasion; now reverse the poles, as before, keeping the electrodes *in situ*,—this makes the electrode over the abrasion, positive, and no pain or irritation will now be felt at this point, but rather a soothing effect. This would seem to prove that there is but one cur-

rent and that it flows in one particular direction, making a positive and negative pole, which act differently
upon both sensory and motor nerves, just as in the
case of a Galvanic battery. Let us now remove the
electrodes and connect the electrophores or cords
to our milli-ampere-meter or Galvanometer, screwing
down the contact screw tight upon our rapid vibrator,
then make and break the primary or battery-circuit of
our induction coil by means of the hand-switch; we
shall observe a throw of the meter-needle in opposite
directions with each make and break of the circuit,
and the degree of deflection will be about the same
for both the make and the break, leading us to con-
clude that a current is generated in the secondary
coil both with the make and the break in the primary
circuit, and that these currents are of about the same
volume, and that they flow in opposite directions,
thus constantly changing the polarity of the coil,
which is, physically speaking, neither positive nor
negative. Both these sets of observations—the
physiological and the physical—are correct, nothwith-
standing their apparent contradiction. One current,
that which is generated upon making the primary
circuit, has so low an electro-motive force that it acts
like one cell of a Galvanic battery, failing to overcome
the *high* resistance of the tissues and cause the flow
of enough current to excite sensory and motor impressions, while it has ample electro-motive force to
set up a sufficient flow of current through the *low*

7 QQ

resistance of our meter to induce a deflection. Therefore, while it is not physically proper to mark the poles of an induction coil as negative (N) and positive (P), it is physiologically proper to mark them as anodal (An) and kathodal (Ka), since such a coil produces but one physiologically active current which flows in one particular direction, and, with the exception that it is intermittent and (as ordinarily produced) of comparatively small volume, presents the same characteristics as an ordinary Galvanic current. The only physical difference between a Galvanic and a Faradic current, consists in the fact that the former may be *continuous*, while the latter is, of necessity, *intermittent*; as regards their respective volumes and electro-motive forces, there is not *necessarily* any difference, since the one may be made to have the same volume and electro-motive force as the other, although as we generally employ these, the Faradic has the lesser volume and greater electro-motive force. If we take two intermittent currents, the one a Faradic and the other a Galvanic, the intermissions of each occuring with the same periodicity, and the electro-motive forces and volumes of each being identical, and apply them in every possible manner to living tissues, we find that the one produces in every instance exactly the same physiological phenomena as the other. We find, then, that the widely varying physiological phenomena which are known to accompany the exhibition of these two so-called different

forms of electricity, are entirely due to the varying
conditions under which one and the same form of
energy is administered, and that these varying conditions have reference to the electro-motive force,
volume, and degree of constancy or intermittency,
of the current. *How important it is, then, to be provided with apparatus and instruments of precision,
for producing and determining these conditions with
absolute accuracy.* In evidence of the chaotic and contradictory notions which prevail concerning this subject, we will quote from Dr. George J. Engelmann,
who is one of the most recent and scientific of writers upon the subject, and this without intending any
uncomplimentary reflection or criticism upon his
work, which has been of a most laudable character,
the profession of America owing him much grati-
tude for his early and scientific presentation of the
modern methods of using electricity in gynæcological practice. On page 37 of his monograph entitled
"The use of Electricity in Gynæcological Practice,"
Dr. Engelmann, in speaking of the Faradic current,
says:

"A difference between the therapeutic effects of
the two poles I have not discovered   *   *   *   *
I shall, however, continue my observations, as I am
by no means fully satisfied as to their identity."
And again on page 19 he says:

" Striking as is the difference between the effect
of the positive and negative poles of the constant

current, diametrically opposed in their chemical
action, I have discovered no difference between the
poles of the Faradic interrupted current, when used
as such, *and it is natural that it should be so, as
with each interruption the current springs from pole
to pole—to and fro."* He, however has observed
and admits the clinical fact that one pole produces
an irritation and pain that is not produced by the
other pole, since he continues by saying: "The negative pole is, however, the more intense and painful."

On page 38 we find this statement: "It is a griev-
ious fault of the Faradic batteries made for medical
purposes that they cannot be sufficiently regulated; the
strength of the current only can be changed, but the
most important feature, the tension of the current, is
fixed." From the explanations and descriptions already given it must be apparent that this is a gross
error. That is just what we do — vary the tension or
electro-motive force of the Faradic current—when
we move the secondary coil to and away from the
primary one. Dr. Engelmann admits that this varies
the "strength" of the current. Now pray tell how is
this accomplished in the face of Ohm's law $C=\frac{E}{R}$ (current equals tension divided by the resistance) without
varying the tension when the resistance remains the
same, as is always the case? This is a mathematical

---

* The author has italicized this portion.

problem the writer would like Dr. Engelmann to solve
for him. Surely, if C changes, we must change the
value of E, when R does not change. We will now
prove the error by a practical experiment, for the
benefit of those who are not sufficiently versed in
mathematics to appreciate the force of the above
reasoning. Attach two electrodes to the secondary
wire of Engelmann's finest or longest coil, place one
electrode in each hand, now move the secondary coil
gradually over the primary one, up to the point where
the current just begins to be appreciable—suppose
this to be 35 mm. from o. Now remove this coil and
place Engelmann's coarsest wire or shortest coil in the
same position, we find that the current is not appreciable. Why? Because the tension or electro-motive
force of the first or longest coil was high enough to
create a flow of current through the high resistance
of the body of sufficient strength or volume to produce a sensory impression, while that electro-motive
force set up by the second or shorter coil when in the
same position with respect to the primary coil, was
not sufficient to set up through the same resistance a
current of similar strength. The fine wire or long
coil, therefore, gives the greater tension, strictly in
accordance with Engelmann's views. But let us now
move the second or short wire coil a little further up
the scale—say to 40 mm. from o, where we find that the
current again becomes appreciable and feels just as
it did when coming from the long wire coil when at

the position marked 35 upon the scale. We have here produced the same increase in the tension of the current by simply moving the coarser secondary coil further up the scale, thus bringing more convolutions or turns of wire into the magnetic field to be cut by lines of force. Moving the shorter wire coil up the scale, is, then, equivalent (up to a certain point) to substituting a longer wire coil, as far as the tension or electro motive force of the current is concerned. The resistance remaining the same, as in the latter experiment, it is a physical impossibility to increase the strength of the current without increasing the tension or electro-motive force; *this is a proposition that can not be gainsaid.*

In another place, upon page 38, Dr. Engelmann says:

"The *tension* of the current is a quality possessed by Faradism in a higher degree than by any other form of electricity."

This is likewise an erroneous idea. Any of the other *so-called* forms of electricity may be caused to have the same electro-motive force (*tension*) that the Faradic current may have, and as we meet with them in medical practice, the *so-called* static machine produces currents of an infinitely higher electro-motive force (*tension*); for instance, the ordinary medical induction coil will only produce an electro-motive force (*tension*) of from 50 to 100 volts, at the outside, while the *so-called* static current has an electro-motive

force of 60,000 volts and upwards, and it can be used
for the same purposes as the Faradic current, and
under appropriate conditions may serve as the most
powerful muscle contracting agent that we have at our
command, notwithstanding its inappreciable volume
even on a "*short-circuit*."

One cell of the new double-cylinder "Law" battery is sufficient to operate the Du Bois-Raymond induction coil shown in Fig. 22, and this will prove the
best type of cell for this purpose.

THE FARADOMETER is an entirely new electrical
measuring instrument, invented by the author for the
measurement of Faradic currents: The term "*Faradometer*" has been coined as most suitable for its desig-
nation. It is a direct-reading, dead-beat instrument;
is not affected by ordinary extraneous inductive in-
fluences, and; does not have to be leveled or placed in
any particular position. It is very sensitive and
reads up to one milli-ampere by hundredths, each of
the smallest divisions indicating one one-hundredth
of one milli-ampere. The need for such an instrument has long been recognized by the most advanced
electro-therapeutists, but until now it has not been
forth-coming. The accompanying illustration, Fig.
24, gives a very clear conception of its external appearance and a fair idea of its internal construction.
A description of the details of its construction and
the principle of its operation would be rather too
technical for the present work. It is placed directly

in circuit (series) with the secondary wire of the in
duction-coil, as shown in Fig. 29, Vol. II, the current being led in at the binding-post A, and out at the
binding-post B, Fig. 24. It may be included in the
circuit or short-circuited (excluded from the circuit)
by withdrawing and inserting a metallic plug into the

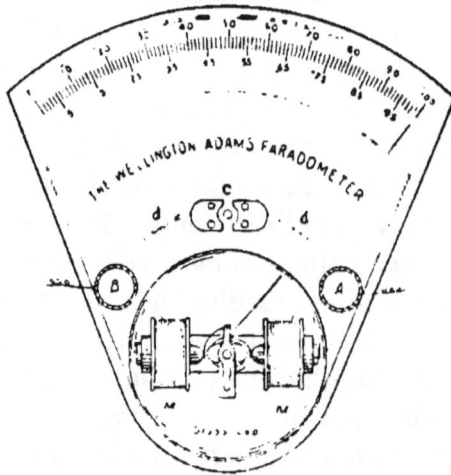

Fig. 24.—The Wellington Adams Faradometer.

hole C; when the plug is in, a *short-circuit* or by-
path is established through d. d., and the current is
thus *"shunted"* or switched around the instrument,
and it does not operate, while when the plug is removed, the current passes through the magnets m. m.,
thus operating the instrument normally to measure the strength or volume of the current being sent
through the tissues, but by adding a suitable resis-

tance and connecting the binding-posts A and B to the terminals of the induction coil, at the same time that a current is being sent through a patient, the *tension* or electro-motive force of the current may likewise be measured directly in volts. With this instrument at our command we can observe and record the qualities—tension and volume—of the Faradic current used upon a patient, in terms of the standard units—the volt and the ampere. The "Weston Electrical Instrument Co." of Newark, N. J., expect to undertake its manufacture.

The introduction of the Faradometer is certain to mark a new era in the history of electro-therapeutics.

INDEX. VOL. I.

8 QQ

BULLETIN OF PUBLICATIONS

– OF –

GEORGE S. DAVIS, Publisher.

THE THERAPEUTIC GAZETTE.

A Monthly Journal of Physiological and Clinical Therapeutics.

EDITED BY

ROBERT MEADE SMITH, M. D.

SUBSCRIPTION PRICE, $2.00 PER YEAR.

THE INDEX MEDICUS.

A Monthly Classified Record of the Current Medical Literature of the World.

COMPILED UNDER THE DIRECTION OF

DR. JOHN S. BILLINGS, Surgeon U. S. A,
and DR. ROBERT FLETCHER, M. R. C. S., Eng.

SUBSCRIPTION PRICE, $10.00 PER YEAR.

THE AMERICAN LANCET.

EDITED BY

LEARTUS CONNOR, M. D.

A MONTHLY JOURNAL DEVOTED TO REGULAR MEDICINE.

SUBSCRIPTION PRICE, $2.00 PER YEAR.

THE MEDICAL AGE.

EDITED BY

B. W. PALMER, A. M., M. D.

A Semi-Monthly Journal of Practical Medicine and Medical News.

SUBSCRIPTION PRICE, $1.00 PER YEAR.

THE WESTERN MEDICAL REPORTER.

EDITED BY

J. E. HARPER, A. M., M. D.

A MONTHLY EPITOME OF MEDICAL PROGRESS.

SUBSCRIPTION PRICE, $1.00 PER YEAR.

THE DRUGGISTS' BULLETIN.

EDITED BY

B. W. PALMER, A. M., M. D.

A Monthly Exponent of Pharmaceutical Progress and News.

SUBSCRIPTION PRICE, $1.00 A YEAR.

New subscribers taking more than one journal, and accompanying subscription by remittance, are entitled to the following special rates.

GAZETTE and AGE, $2.50; GAZETTE, AGE and LANCET, $4.00; LANCET and AGE, $2.50; WESTERN MEDICAL REPORTER or BULLETIN with any of the above at 20 per cent. less than regular rates.

Combined, these journals furnish a complete working library of current medical literature. All the medical news, and full reports of medical progress.

GEO. S. DAVIS, Publisher, Detroit, Mich.

IN EXPLANATION

OF

The Physicians' Leisure Library.

We have made a new departure in the publication of medical books. As you no doubt know, many of the large treatises published, which sell for four or five or more dollars, contain much irrelevant matter of no practical value to the physician, and their high price makes it often impossible for the average practitioner to purchase anything like a complete library.

Believing that short practical treatises, prepared by well known authors, containing the gist of what they had to say regarding the treatment of diseases commonly met with, and of which they had made a special study, sold at a small price, would be welcomed by the majority of the profession, we have arranged for the publication of such a series, calling it **The Physicians' Leisure Library.**

This series has met with the approval and appreciation of the medical profession, and we shall continue to issue in it books by eminent authors of this country and Europe, covering the best modern treatment of prevalent diseases.

The series will certainly afford practitioners and students an opportunity never before presented for obtaining a working library of books by the best authors at a price which places them within the reach of all. The books are amply illustrated, and issued in attractive form.

They may be had bound, either in durable paper covers at **25 Cts.** per copy, or in cloth at **50 Cts.** per copy. Complete series of 12 books in sets as announced, at **$2.50,** in paper, or cloth at **$5.00,** postage prepaid. See complete list.

PHYSICIANS' LEISURE LIBRARY

PRICE: PAPER, 25 CTS. PER COPY, $2.50 PER SET; CLOTH, 50 CTS. PER COPY, $5.00 PER SET.

SERIES I.

Inhalers, Inhalations and Inhalants.
By Beverley Robinson, M. D.

The Use of Electricity in the Removal of Superfluous Hair and the Treatment of Various Facial Blemishes.
By Geo. Henry Fox, M. D.

New Medications, Vol. I.
By Dujardin-Beaumetz, M. D.

New Medications, Vol. II.
By Dujardin-Beaumetz, M. D.

The Modern Treatment of Ear Diseases.
By Samuel Sexton, M. D.

The Modern Treatment of Eczema.
By Henry G. Piffard, M. D.

Antiseptic Midwifery.
By Henry J. Garrigues, M. D.

On the Determination of the Necessity for Wearing Glasses.
By D. B. St. John Roosa, M. D.

The Physiological, Pathological and Therapeutic Effects of Compressed Air.
By Andrew H. Smith, M. D.

Granular Lids and Contagious Ophthalmia.
By W. F. Mittendorf, M. D.

Practical Bacteriology.
By Thomas E. Satterthwaite, M. D.

Pregnancy, Parturition, the Puerperal State and their Complications.
By Paul F. Mundé, M. D.

SERIES II.

The Diagnosis and Treatment of Haemorrhoids.
By Chas. B. Kelsey, M. D.

Diseases of the Heart, Vol. I.
By Dujardin-Beaumetz, M. D.

Diceases of the Heart, Vol. II.
By Dujardin-Beaumetz M. D.

The Modern Treatment of Diarrhoea and Dysentery.
By A. B. Palmer, M. D.

Intestinal Diseases of Children, Vol. I.
By A. Jacobi, M. D.

Intestinal Diseases of Children, Vol. II.
By A. Jacobi. M.

The Modern Treatment of Headaches.
By Allan McLane Hamilton, M. D.

The Modern Treatment of Pleurisy and Pneumonia.
By G. M. Garland, M. D.

Diseases of the Male Urethra.
By Fessenden N. Otis, M. D.

The Disorders of Menstruation.
By Edward W Jenks, M. D.

The Infectious Diseases, Vol. I.
By Karl Liebermeister.

The Infectious Diseases, Vol. II.
ry Karl Liebermeister.

SERIES III.

Abdominal Surgery.
By Hal C. Wyman, M. D.

Diseases of the Liver
By L ujardin-Beaumetz, M. D.

Hysteria and Epilepsy.
By J. Leonard Corning, M D.

Diseases of the Kidney.
By Dujardin-Beaumetz, M. D.

The Theory and Practice of the Ophthalmoscope.
By J. Herbert Claiborne, Jr., M D.

Modern Treatment of Bright's Disease.
By Alfred L. Loomis, M. D.

Clinical Lectures on Certain Diseases or Nervous System.
By Prof. J. M Charcot, M. D.

The Radical Cure of Hernia.
By Henry O Marcy, A M., M. D., L. L D.

Spinal Irritati n.
By William A Hammond, M. D.

Dyspepsia.
By Frank Woodbury, M. D.

The Treatment of the Morphia Habit.
By Erlenmeyer

The Etiology, Diagnosis and Therapy of Tuberculosis
By Prof. H. von Ziemssen.

SERIES IV.

Nervous Syphilis.
By H. C. Wood, M. D.

Education and Culture as correlated to the Health and Diseases of Women.
By A. J. C. Skene, M. D.

Diabetes.
By A. H. Smith, M D.

A Treatise on Fractures.
By Armand Després, M. D.

Some Major and Minor Fallacies concerning Syphilis.
By E. L. Keyes, M. D.

Hypodermic Medication.
By Bourneville and Bricon.

Practical Points in the Management of Diseases of Children.
By I. N. Love, M. D.

Neuralgia.
By E. P. Hurd, M. D.

Rheumatism and Gout.
By F. Le Roy Satterlee, M. D

Electricity, Its Application in Medicine.
By Wellington Adams, M.D. [Vol.I]

Electricity, Its Application In Medicine.
By Wellington Adams, M.D. [Vol.II]

Auscultation and Percussion.
By Frederick C. Shattuck, M. D.

SERIES V.

Taking Cold.
By F. W. Bosworth, M. D.

Practical Notes on Urinary Analysis.
By William B. Canfield, M. D.

Practical Intestinal Surgery. Vol. I.
Practical Intestinal Surgery. Vol. II.
By F. B. Robinson, M. D.

Lectures on Tumors.
By John B. Hamilton, M. D., LL. D.

Pulmonary Consumption, a Nervous Disease.
By Thomas J. Mays, M.D.

Lessons in the Diagnosis and Treatment of Eye Diseases.
By Casey A. Wood, M. D.

Diseases of the Bladder and Prostate.
By Hal C. Wyman, M. D.

Artificial Anæsthesia and Anæsthetics.
By DeForest Willard, M. D., and Dr. Lewis H. Adler, Jr.

Cancer.
By Daniel Lewis, M. D.

The Modern Treatment of Hip Disease.
By Charles F. Stillman, M. D.

Insomnia and Hypnotics.
By Germain Seé.
Translated by E. P. Hurd, M. D·

BOOKS BY LEADING AUTHORS.

SEXUAL IMPOTENCE IN MALE AND FEMALE $3.00
By Wm. A. Hammond, M. D.
PHYSICIANS' PERFECT VISITING LIST 1.50
By G. Archie Stockwell, M. D.
A NEW TREATMENT OF CHRONIC METRITIS .... .50
By Dr. Georges Apostoli.
CLINICAL THERAPEUTICS.......................... ...... 4.00
By Dujardin-Beaumetz, M. D.
MICROSCOPICAL DIAGNOSIS........................... 4.00
By Prof. Chas. H. Stowell, M. S.
PALATABLE PRESCRIBING. 1.00
By B. W. Palmer, A. M., M. D.
UNTOWARD EFFECTS OF DRUGS 2.00
By L. Lewin, M. D.
SANITARY SUGGESTIONS (Paper)....................... .25
By B. W. Palmer, M. D.
SELECT EXTRA-TROPICAL PLANTS.... .... 3.00
By Baron Ferd. von Mueller,
TABLES FOR DOCTOR AND DRUGGIST 2.00
By Eli H. Long, M. D.

GEORGE S. DAVIS, Publisher,

P O. Box 470   Detroit, Mich.